Second Nature

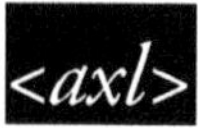

Axl Books
Box 19009
104 32 Stockholm
Sweden
www.axlbooks.com
info@axlbooks.com

ISBN 978-91-978598-8-2

SECOND NATURE
Origins and Originality in Art, Science, and New Media

Rolf Hughes
Jenny Sundén
(EDS.)

AXL BOOKS

Contents

Acknowledgements

Editing a book is always an effort that draws on many diverse contributions, and there are accordingly many people to whom we are indebted. First and foremost, we thank the authors, whose talent, care, and, not least, patience helped bring this book into being. It has been a pleasure working with you. We also thank Staffan Lundgren, our publisher at Axl Books. You saw a book from the moment the manuscript landed on your desk – for this, and for the constructive work that followed this initial judgement, we are truly grateful. Last, but not least, we acknowledge our lasting gratitude to the Swedish Research Council, which has made this book possible by providing funding for the research projects "Technologies of Life: On Art, Science and New Media" (Sundén) and "Auto-poiesis and Design: Authorship and Generative Systems" (Hughes).

Rolf Hughes and Jenny Sundén
Stockholm, May 2011

Foreword

Jay David Bolter

> We define the aura [of natural objects] as the unique phenomenon of a distance, however close it may be. If, while resting on a summer afternoon, you follow with your eyes a mountain range on the horizon or a branch which casts its shadow over you, you experience [breathe] the aura of those mountains, of that branch.[1]

It should not be surprising that an anthology devoted to the relationship of nature and technology should cite Walter Benjamin. The idea that reproducibility leads to a loss of aura in the technologies of photography and film has had lasting resonance not only for media theory but also for cultural theory in general. In the passage above, from "The Work of Art in the Age of Mechanical Reproduction," Benjamin describes the experience of aura in traditional art by appealing to nature. He claims that we can feel a special relationship to natural objects such as mountains or trees, presumably because of they are uniquely situated and present to us. It is this presence and uniqueness that traditional works of art also possess and that the new reproductive technologies destroy.

Benjamin's description of aura here could almost be mistaken for a Romantic sense of union with nature, except for the operative clause in his definition: "the unique phenomenon of a distance, however close it may be." Benjamin claims that we feel the aura of natural objects precisely when we are aware of our separation from them. We also feel this sense of separation in the case of art objects produced through the traditional technologies of painting or sculpture, because the uniqueness in time and space of the objects constitutes an inseparable barrier. We realize that although we could touch these objects or even deface them, we can never touch or alter their history. By destroying the uniqueness of the art object, reproductive technologies such as film and photography make art available to the masses for appropriation and critique: they diminish the distance between art and the spectator. Ben-

1. Walter Benjamin, "The Work of Art in the Age of Mechanical Reproduction," in *Illuminations*, trans. Harry Zohn (New York: Schocken Books, 1968), 222-223.

jamin suggests, for example, that in cinema the camera penetrates and dissects the space of the dramatic action and even of the film actor himself.

In our age of biotechnology, Benjamin's argument does seem to be as appropriate to nature as to art. Biotechnology can be read as an assault on the aura that was ascribed to nature in the nineteenth-century and that Benjamin still seems to recall. The complexities and subtleties of that assault constitute the subject of this anthology, as its authors explore the ways in which the boundary between nature and technology is now dissolving – in particular as they consider how the human body is serving as a site for that dissolution. The editors, Rolf Hughes and Jenny Sundén, are certainly aware of Benjamin's analogy between art and nature, and the interpenetration of nature by technology that the analogy suggests.

The title of their volume, *Second Nature*, prompts us as readers to question whether there is such a thing as a first nature, a state of nature prior to technology. In our current cultural moment, the editors ask, doesn't nature always come to us already technologized, already mediated? An original, untouched, unmediated nature would be auratic, just as a work of art is auratic that has a unique and secure origin in time and space. Other contributors also make explicit reference to Benjamin. Boo Chapple reverses Benjamin's analogy, when she suggests that the reproductive technologies of film and photography could now serve as models for the controlled reproduction (cloning and genetic modification) in the biotech industries. Maria Chatzichristodoulou invokes Benjamin to argue that in order to appreciate performance art mediated through networked technologies we need a more nuanced notion of presence and embodiment: Benjamin's conflation of aura, uniqueness, and presence no longer suffices. Benjamin's notion of reproduction only provides one approach to this fascinating anthology. Most of the authors do not directly invoke Benjamin, but I think that many of their approaches could be seen as versions of Benjamin's trope of "distance-no-matter-how-near," where the question is whether the traditionally assumed distance between nature (or the animal or the body) and technology can be maintained in the face of biotechnology and its cultural practices.

This is not to say that we can adopt Benjamin's analysis without reservation – either for media theory or for the study of biotechnology. It could be argued that even in the age of mechanical reproduction, our culture kept trying to reinscribe aura in film and other popular media forms. Hollywood attempted to institutionalize aura, a fact that Benjamin grudgingly acknowledged when he condemned the false aura, or cult of personality, of the star system. As I have argued elsewhere, our culture continues to reinscribe aura in certain entertainment forms today.[2] Not only in popular film but to some extent in television and even computer games, we are witnessing not the final destruction of aura, but a perpetual crisis of aura, in which authenticity and origin are called into question, reasserted, and then questioned again. In this context media producers seek to manufacture aura through reproduction. If in traditional art, at least for Benjamin, aura was an inherent quality, in contemporary media forms, aura becomes a design parameter.

Attempts to reinscribe aura may redouble with the development of new media. The digital production of images should be the ultimate vindication of Benjamin, the ultimate loss of aura. If the process of technical reproduction created the crisis of aura, then digital copying reenacts and intensifies that crisis in every single moment of digital production. Yet, digital technology also undermines the very notion of the original, at least for works that are created in the new technology. If an artist uses graphics software to create an image, then she is making copies throughout the process of creation: the original is already the product of dozens or hundreds of copying operations. And if the new media designer can reinscribe aura, what about the biotechnician?

As W J T Mitchell has pointed out in "The Work of Art in the Age of Biocybernetic Reproduction," biotechnology can be seen as the combination of digital media technology and biology. The use of computers to chart and control the replication of genetic material assimilates the work of the biotechnician to that of the new media designer and makes Benjamin's analysis of the loss of

2. Jay David Bolter et.al., "New Media and the Permanent Crisis of Aura," *Convergence* 12, 1 (2006).

aura particularly relevant for biotech. Mitchell also suggests, however, that Benjamin's argument about the loss of aura must be rethought:

> Biocybernetic reproduction carries [the] displacement of the original one step further, and in doing so reverses the relation of the copy to the original. Now we have to say that the copy has, if anything even more aura than the original...[I]f aura means recovering the original vitality, literally the 'breath' of life of the original, then the digital copy can come closer to looking and sounding like the original than the original itself.[3]

The editors and authors of this anthology are engaged in just such a rethinking. They are rethinking the relationship of copy to original; they are revisiting the "distance-no-matter-how-near" that now separates our biotechnological second nature from a supposed first nature.

3. W. J. T. Mitchell, *What Do Pictures Want? The Lives and Loves of Images* (Chicago: University of Chicago Press, 2005), 319-320.

Nature Seconded

Jenny Sundén
Rolf Hughes

All our service
In every point twice done and then done double.
Macbeth, I.vi

I believe that animals are descended from at most only four or five progenitors, and plants from an equal or lesser number. Analogy would lead me one step further, namely, to the belief that all animals and plants are descended from some one prototype. But analogy may be a deceitful guide.
Charles Darwin[1]

There is no origin of species because there is no unity from which descent is derived, only types, variations of differences and types of reproduction and descent, both of which must be assumed from the start.
Elizabeth Grosz[2]

The urge to conceive not only a new art or life form, but a new self or subject worthy of them, is an enduring one. In modernism, enigmatic questions such as *Where do we come from? Who are we? Where are we going?* became not only, as Hal Foster reports, "a great catechism" for artists, but persist even today, allowing artists and others to ask questions of first and last things – "the beginnings of subjectivity and sexuality, the ends of art and imagination."[3] Questioning origins and originality, during a period characterized by rapid convergences of the tools, methods and terminologies of art, science and technology, raises some of the most pressing questions in contemporary research, such as issues of agency and accountability, hybridity and identity, intellectual property and oeuvre, intention and authority. These, and a constellation of related philosophical, economic, aesthetic, legislative and political concerns inseparable from historical paradigms of origins and originality, are today subjected to rapid reconfiguration due to the current pace of technological and theoretical change. The notion of an

1. Charles Darwin, *The Origin of Species* (Oxford: Oxford University Press, 1859/1996), 642.
2. Elizabeth Grosz, *The Nick of Time: Politics, Evolution, and the Untimely* (Durham and London: Duke University Press, 2004), 25.
3. Hal Foster, *Prosthetic Gods* (Cambridge, MA: MIT Press, 2004), ix.

autonomous, responsive, actively reproductive machine or artificial "intelligence" – not to mention the hybrid fusions of "bioart," some of which are discussed in this book – destabilize assumptions underpinning modern concepts of self and originality, as well as established distinctions between human and non-human, distinctions that can be mediated by another suffix, one simultaneously convenient and provocative – namely, the *post*-human.[4]

Several essays in this collection focus on questions of embodiment, technology, and reproduction (see, for example, Ron Broglio, Maria Chatzichristodoulou, Francois-Joseph Lapointe & Martine Époque, and Jenny Sundén). Some explore the idea of embodiment as being circulated or distributed over human and non-human components. Echoing Gail Weiss's notion of "intercorporeality," (human) bodies in this sense are conceived not as discrete and autonomous entities, but as *experiences* of being embodied, experiences that are understood as intersubjective, materializing out of embodied relationships and connections that are continuously altering them.[5] In her work on technobodies, Weiss' notion of intercorporeality takes in not only the interconnectedness between human bodies, but between bodies and body-parts (human and non-human) – i.e. between bodies and technological practices that are persistently transforming and redefining them.[6] How do technologies of reproduction, in art and in science, alter the interface between bodies and machines, nature and culture? How is hardware, software, and wetware configured in reproduction machines? If human reproduction is increasingly technologized, we need to rethink what is meant by being born. Where does our story begin? As Donna Haraway's emblematic cyborg figure makes clear, the merging of the technological with the organic calls for alternate ways of conceptualizing embodied subjectivity.[7]

4. On the post-human, see, for example, N. Katherine Hayles, *How We Became Posthuman: Virtual Bodies in Cybernetics, Literature, and Informatics* (Chicago and London: Chicago University Press, 1999).
5. Gail Weiss, *Body Images: Embodiment as Intercorporeality* (London and New York: Routledge, 1999).
6. Gail Weiss, "The Durée of the Technobody," in *Becomings: Explorations in Time, Memory, and Futures*, ed. Elizabeth Grosz (Ithaca and London: Cornell University Press, 1999).
7. See Donna Haraway, "A Cyborg Manifesto: Science, Technology, and Socialist-Feminism in the Late Twentieth Century," in *Simians,*

Other contributions also call into question our understanding of embodied subjectivity from different perspectives, examining how changing paradigms of cultural production impact on our relation to creativity, individuality and uniqueness (see, for example, Boo Chapple, Maria Chatzichristodoulou, Rolf Hughes, and Timothy Weaver). By investigating design practices as material processes that make possible new kinds of alignments and affinities between author, technology, and artifact or experience, such studies frame a range of questions of broad cultural and epistemological significance: Is human judgment, informed by the supposedly reliable criteria of familiarity with the history of conventions within a given practice, still the core criteria for evaluating the quality and authenticity of a work? How do prevailing metaphors of reproduction influence the research logic of the sciences, humanities and interdisciplinary forms of inquiry? What is the basis of our potentially fatal attraction to both conceptual and technological innovation and the copy, the virus, the sample, and the clone? What is involved in the shift from a traditional economy of art in imitation of nature, to one that subsumes or incorporates nature self-reflexively? Which is to say how might we appropriately position ourselves in relation to an artistic medium if the artistic medium itself is *alive*?

Historically, the notion of "second" nature has assumed a distinction between nature and culture – that is, between the pre-existing (animal, mineral, vegetable) and the wrought (language, images, spectacle, sound and performance). This implies in turn a theory of time and thus of *origins*. The contours of the Western imagination has been defined by two primary accounts of how the world (or, let's simply say, *everything*) originates.[8] In the pagan tradition, the account presented in Plato's *Timaeus*, the creator gives pre-existing form to pre-existing matter and thereby sets time in motion and, with it, the possibility of change. In the Judeo-Christian account, however, the God of Augustine's *Confessions*

Cyborgs and Women: The Reinvention of Nature (New York: Routledge, 1991).

8. This account of the pagan and Judeo-Christian origin narratives is drawn from Richard Shiff's "Originality," in *Critical Terms for Art History*, ed. Robert S. Nelson and Richard Shiff (Chicago: University of Chicago Press, 2003), 145-159.

is more *original*, creating "something, and out of nothing" (and thereby disproving Lear's admonition to Cordelia in Act I Scene i of *King Lear*, that "Nothing will come of nothing.") In both accounts, history is seen to need time and a certain "something," a material basis for growth and decay. Artists arrive at this particular party somewhat belatedly, playing neither the role of pagan prime mover nor of Christian Prime Innovator (God); rather they take up what already exists (the something) and thereby enter into dialogue with historical traditions, with a pre-existing language of form. This distinction informs the mythologies that have developed around originality. As Shiff remarks, originality implies an order, an origin – a priority or lack of precedent – and therefore involves considerations of chronology and historical sequence, as well as class issues, social priority or lineage:

> If artists must use what has already been shaped, how can they and their artworks attain originality? Perhaps originality is transmissible (the artist as inheritor and bearer of original first principles, a set of universal truths). And perhaps originality is manifested when one alters existing directions or forces (the artist as counter-cultural deviator of a tradition or as social deviant).[9]

What is it that links artistic originality (be that a *something from nothing*, or a *deviation* from tradition) and second nature (with its connotation of *alterity* – second nature as a step away from a supposedly more *natural* original nature)?[10] A first answer would be the notion of *poiesis*, the concept of making and creating (or intervening and designing).[11] Poiesis brings into play human agency

9. Shiff, *Critical Terms for Art History*, 45.
10. From the Renaissance, interest shifted from understanding nature to understanding *and* controlling it. Bacon proclaimed in 1620 that we command nature only by submitting to it. In the 17th and early 18th century originality was closely linked to the Aristotelian conception of imitation (*mimesis*) of an idealized, external nature, but from the mid 18th century the focus shifts to interior, emotional and subjective states and the images and representations of artistic expression itself. Individual creativity increasingly supplants the mapping of one mimetic world onto another as the subject of images and representations.
11. In the Symposium, Plato has Diotima remark "There is poetry, which, as you know, is complex and manifold. All creation or passage of non-

– the imagination articulated through specific skills and tools, all historically and culturally bound. The practices of new media, for example, typically involve different interactions – between human agency and materials/tools, design team and client, architect and engineer – and thus generate new types of authorship, relationships between producers and consumers, and distribution models, thereby acting, as Lev Manovich has argued, as the avant-garde of the culture industry[12]. In the West, it is the ability to create or design something *deliberately* and *at will* that usually denotes creative competence (or confers the status of author, artist, designer etc. on the person acting in such a way) – an accidental result is not usually counted as a legitimate creative act, product, or artifact – which implies that such skills can be taught, passed on by example or by deliberate instruction. The notion of an "autonomous," "creative" designing intelligence in a machine profoundly unsettles such assumptions. A concept such as that of *artificial intelligence* thus calls for clarification of the notions of *making* and *artifice* as opposed to *natural* or *chance* production. A product with an identifiable authorial origin (or set of origins, as in the case of design teams working on component parts) has been made for a purpose; if it is well designed it will serve that purpose (the designer's purpose), thus its performance in serving that purpose is the designer's responsibility. An intelligent being (artificially intelligent or otherwise) would need to have its own purposes and projects and be therefore responsible for its own actions. But (as Mary Shelley's *Frankenstein* nicely illustrates) a machine cannot be responsible independently of its creator. Even programming randomizing elements to create unpredictable behavior does not provide evidence of a separate will or point of view on the part of the machine – rather, it simply makes the machine embody the programmer's exact intentions

being into being is poetry or making, and the processes of all art are creative; and the masters of arts are all poets or makers. [...] [T]hey are not called poets, but have other names; only that portion of the art which is separated off from the rest, and is concerned with music and metre, is termed poetry, and they who possess poetry in this sense of the word are called poets." Plato, *Symposium*, trans. Benjamin Jowett (Adelaide: University of Adelaide, 2004), accessed June 21, 2010, http://ebooks.adelaide.edu.au/p/plato/p71sy/symposium.html.

12. See Lev Manovich, *The Language of New Media* (Cambridge, MA: MIT Press, 2002).

(creating unpredictability). This suggests that a closer look at that most unruly of human qualities – the imagination – might usefully inform our inquiry.

Richard Kearney, in his study of the formative concepts of human imagination, identifies the main Hebraic term for imagination as *yetser*, a word derived from the same root (*yzr*) as the terms for "creation" (*yetsirah*), "creator" and "create" (*yatsar*), and argues that such "allusive interplay between the terms used to describe God's creation of the world and the First Man's transgressive capacity (i.e. the *Yetser*) to imitate this divine act is highly significant. When God "created (*Yatsar*) Adam in his own image (*tselem*) and likeness (*demuth*)," (Gen. 2:8) He risked allowing man [*sic*] to emulate Him, to set himself up as His rival, to supplant Him in the order of creation."[13] Thus if we understand the *yetser* as humankind's creative impulse to imitate God's own creation, then the moment when Adam and Eve ate the forbidden fruit of the Tree of Knowledge represents its first instance of enactment, a moment that simultaneously provides *ethical* knowledge of good and evil and *historical* knowledge of past and future:

> Imagination enables man to think in terms of opposites – good and evil, past and future, God and man. Thus bringing about the consciousness of sin and of time, the fallen imagination exposes man to the experience of division, discord and contradiction.[14]

The Hebraic imagination is thus simultaneously a liberation and a curse – a means of conceiving of the majesty of God but also of rebelling against such majesty primarily through erotic impulses and sexual desire – i.e. the ungovernable imagination. Yet if desire here is focused on Eve, then its object is one who is *made* not born – Eve, in fact, incarnates the adaptive forces at work in originality, linking origin indissolubly to artifice. Human nature and second nature are thus bound together in a symbolic network of desire and counter-desire, nature and reification – the authentic and the

13. Richard Kearney, *The Wake of Imagination* (Minneapolis: University of Minnesota Press, 1988), 39.
14. Kearney, *The Wake of Imagination*, 40.

fake clasped together in an indissoluble embrace.[15]

From the outset then, we confront the disruptive ambiguity of a human imagination created by God, but linked to rebellion and therefore being the responsibility of humankind to contain and control.[16] In such an account of origins, humankind, after the Fall but not yet manufacturing artificial paradises, is already a second nature, wrought by a God who would come to repent endowing His creation with imagination, with *yester*, and for good reason; condemned thereby to rebel against its Creator, and shackled to the idolatrous drives of the imagination, the desire to represent God in our own "graven images," humankind is thus powerless to resist the compulsion to create culture.[17] The Talmudic injunction

15. Marina Warner writes, "At the heart of this web of symbols, where women as *fons et origo* and woman as manufactured maiden are assimilated, where women as original matter and women as artifact become interchangeable terms in the discussion of the creative act in life and art, we can find the source of the tradition of describing meaning more readily to the female form than to the male. The female was perceived to be a vehicle of attributed meaning at the very beginning of the world, according to the myths that lie at the foundation of our lives, ever since she was made in all her allure as man's fatal partner." Marina Warner, *Monuments and Maidens: The Allegory of the Female Form* (London: Weidenfeld and Nicolson, 1985), 224-225.
16. This apparent paradox is examined by Frank Porter in his analysis of the Jewish doctrine of sin, a doctrine based on the two "apparently contrary conceptions" at play in the word *yezer*: "The verb *yzr* means to form, or fashion, and also, to form inwardly, to plan. It was used as the technical word for the potter's work. It was frequently used of God's forming of nature and of man, and also of his planning or purposing. The *yzr* of man could therefore suggest either his form, as God made him, his nature (so *Ps.* 103:14), or his own formation of thought and purpose, "imagination" as the word is rendered in several Old Testament passages (*Gen.* 6.5; 8:21; *Deut.* 31:21; *Isa.* 26:3; I *Chr.* 28:9; 29:18)." See Frank Porter, "The Yetser Hara: A Study of the Jewish Doctrine of Sin," in *Biblical and Semitic Studies* (New York: Scribner, 1901), 109. Cited in Kearney, *The Wake of Imagination*, p. 46.
17. For a different emphasis, see the philosopher and radical theologian Don Cupitt, who writes "In Latin the verb *colo, colere, colui, cultum* means "to till the ground, to tend and care for."" In antiquity it had already gathered the wide range of extended senses it still has today. Culture is the familiarized, tamed, gardened version of the world, which we establish around ourselves. Cicero calls it a "Second Nature." It includes not only the *cultivation* of the soil, *agriculture*, but also the *culture* of new varieties of domesticated plant, the careful *cultivation* of acquaintanceships and skills, of one's own "person" and one's own soul, and therefore also *high culture*, the arts and sciences, and above all the *cult*, the care and tending of the gods." Thus, he continues, our entire range of cultural activities are linked to "a series of symbolic transformations of the activity of the poor plough-

against idolatry, which is explicitly linked to a lack of control over the imagination, results in an *invisible* God, and therefore, as Freud notes, the subordination of sensory perception to an abstract idea – "a triumph of spirituality over the senses; more precisely a renunciation of the satisfaction of an impulse derived from an instinct (*Triebverzicht*)."[18] The bodily and the abstract thus appear as further nodes on our increasingly rhizomatic map of distinctions that multiply from our original theme of "second nature." Furthermore, the interchangeability of woman as origin and woman as, in Marina Warner's felicitous phrase, "manufactured maiden" introduces:

> a paradigmatic metaphor for the act of artistic creation, so that artists "give birth" to their works. These mythological principles, confusing women and art, together underpin the idea that man is a maker and woman made, in mythic reversal of biology (both sexes issuing from woman). They have assisted the projection of immaterial concepts on the female form, in both rhetoric and iconography. Men act as individuals, and women bear the burden of their dreams.[19]

In the Hellenic concept of imagination, we find a counterpart to the biblical story of Adam's fallen imagination. The Greek myth of Prometheus is a pre-philosophical account of humankind's acquisition of the power to shape their world (through the fire that Prometheus stole from the gods), to create words, images, artifacts and performances capable of transforming nature (the cosmos of blind necessity presided over by Zeus) into culture (a realm of relative freedom that permits a measure of human autonomy). Kearney suggests that a comparison of the two cultures reveals a shift in emphasis from the *ethical* to the *epistemological* dimension of imagination. He notes that the name Prometheus means fore-sight (*pro-metheus*) and thus designates "the power to anticipate the future by projecting an horizon of imaginary possibilities." Imagina-

man tilling the soil." See Don Cupitt, *After God: The Future of Religion* (London: Weidenfeld and Nicholson, 1997), 22-23.

18. Sigmund Freud, *Moses and Monotheism* (New York: Vintage, 1939), 144.
19. Warner, *Monuments and Maidens*, 239.

tion is again tainted at the outset by an act of transgression – in this case deception rather than temptation – and is associated with foresight, which permits humanity to imitate the gods. This again marks imagination with an essential ambiguity – a rebellion (the unlawful act of stealing the gods' fire) gives humankind the capacity to imitate God, and thus (in the words of Denis Donoghue):

> the imaginative enhancement of experience, the metaphysical distinction between what happens to us and what we make of this happening. That is to say, Prometheus provided men [*sic*] with consciousness as the transformation grammar of experience. No wonder the gift also gave men a sense of the endlessness of possibility arising from the endlessness of knowledge and desire. The power of (imagination) helped men to maintain a relation between themselves and nature, but it did not bring peace between men and gods.[20]

Essentially contested meanings – not least divinity and mortality, obedience and rebellion – are contested most keenly through, once again, the figure of the woman. As the expression of Zeus' displeasure at Prometheus and agent of divine punishment, she now represents:

> the first imitation, who, replying to the first deception, embodies now for all time the principle of deception. She imitates both divine and bestial traits, endowed by the gods with an exterior of wondrous beauty and adornment that conceals the greedy and thievish nature of her interior. Artifact and artifice herself, Pandora installs the woman as *eidolon* in the frame of human culture, equipped by her "unnatural" nature to delight and deceive.[21]

20. Denis Donoghue, *Thieves of Fire* (London: Faber, 1973), 26. Cited in Kearney, *The Wake of Imagination*, 80.
21. Froma I Zeitlin, "Travesties of Gender and Genre in Aristophanes' 'Thesmophoriazousae,'" in *Reflections of Women in Antiquity*, Helene B. Foley ed. (New York: Gordon and Breach, 1981), 154.

Second can denote not only a sequence, an order(ing), but also a repeat, an attempt to repair a deficiency (second draft, second attempt etc.)[22] According to Plato, Prometheus' theft of fire is an indication that the human "art of making" (*demiourgike techne*) is premised on such a lack (the absence of, and need for, fire) – an acknowledgement of the deficiency of those who need fire to achieve a more perfect form of being. "In obtaining this higher form of being for man," Plato writes in Protagoras, "Prometheus shows himself to be man's double, an eternal image of man's basically imperfect form of being." Once again, imagination is simultaneously liberating, compensating for humankind's sense of lack, and an affliction, an offence against the gods, tainting all generations with the original transgression of Adam or Prometheus. Yet both narratives are also ambiguous, seeing such transgression as the precondition of human culture and imagination as:

> a power which supplements the human experience of insufficiency and sets man up as an original creator in his own right. But because imagination deals in the realm of art rather than nature, it can never fully escape the feeling that it is merely an *imitation* of the *original* act of a divine maker (e.g. the biblical Yahweh or the Greek Demiurge) – an act which alone is deemed lawful. Imagination can never forget that its art is artifice, that its freedom is arbitrary, that its originality is a simulation, repetition, *mimesis*.[23]

These sketches of the separation of humankind from the divine order not only illustrate the historical identification of the human capacity to imagine with transgression, but also reveal in emerging paradigms of nature and the natural, an emulation of (or even

22. "Coming next after the first according to any contentually understood principle of enumeration (e.g. in order of time, position, rank, quality, conventional or arbitrarily adopted sequence)" - the Oxford English Dictionary definition for "second" includes this interesting inversion of Schiff's linking (cited earlier) of originality to a privileged order or sequence. *Oxford English Dictionary*, Second Edition (Oxford: Oxford University Press, 1989), 825.
23. Kearney, *The Wake of Imagination*, 81-82.

competition with) the role of divine beings who (in both Hebraic and Hellenic traditions) shape the world from formless matter, making our world already and always second nature. Second nature is written into our very DNA, it seems. A comparison of Pandora, the first "manufactured maiden" of classical myth, and Eve, the "mother of the living" in the Judeo-Christian story, illuminates important aspects of our shared legacy (as Marina Warner writes); "the fashioned woman who is born beautiful to behold in her future partner's eyes is blamed for the distress that befalls humanity almost immediately after her advent." Furthermore, in both Genesis and Hesiod, her advent and the fall she unwittingly brings about result in monogamous marriage being established as the legitimizing origin of children.[24]

Understanding nature primarily as a shifting classificatory process, Sarah Franklin, Celia Lury, and Jackie Stacey investigate the concept *second nature* partly in relation to the transitive verb "to second": to attend or assist; to promote or reinforce; to transfer or replace.[25] The first sense of to second (to attend or assist) points at notions of "assisted nature" (as in, for example, assisted reproduction). Here nature is remade or reconfigured through technology. The second sense (to promote or reinforce) accentuates uses of the natural as reinforcement of culture. Nature as "seconded" by culture here refers to processes in which nature is used as means of legitimization, or as foundational essentialism. Finally, the authors depict second nature in terms of transfer or replacement, as when qualities of nature are transferred onto culture, or, through a reversal of qualities, when the natural takes the place of the cultural.

In contemporary discourses of reproductive technologies (whether this means artificial insemination, *in vitro* fertilization, "artificial womb" technologies, or something else), the term "artificial" appears to assist and even reinforce the modes and codes of heterosexuality and nuclear family formations (cf. the chapters of Sundén and Wagner). While potentially disruptive of supposedly "natural" ways of living and conceiving (by dis-locating reproductive materials from reproductive bodies), abundant legisla-

24. Warner, *Monuments and Maidens*, 220.
25. Sarah Franklin, Celia Lury and Jackie Stacey, *Global Nature, Global Culture* (London: Sage, 2000), 22-23.

tion exists to ensure the opposite occurs, securing the continuing privilege of heterosexual, married couples by allowing them easier access to technologies of reproduction. Technologies of reproduction simultaneously work to change and control the domain of parenting and motherhood. Mary Ann Doane points out – in an interesting cross-examination of human reproduction as that which organizes generations, whereas mechanical reproduction, such as photography, structures memories – that technologies of reproduction work in a sense to control the excesses of the maternal. When this control is exercised, reproductive technologies do not only make possible new parental definitions and practices, she argues, but also weaken the more positive connotations of the maternal: "Without her, the story of origins vacillates, narrative vacillates. It is as though the association with a body were the only way to stabilize reproduction."[26]

The question is what happens to this epistemological certainty (the fact that the mother is knowable whereas the father is unproven until genetically identified) if such a conception of motherhood also starts to vacillate, not to say duplicate. If an association with a specific body is the only way of stabilizing reproductive anxieties, is such stabilization possible when "mother" can be many? Who counts as the origin, the culturally privileged mother-figure, within paradigms of biotechnological network mothering? Again we return to the question of origins: where does our story begin?

The meaning of second nature developed within the Frankfurt School – by thinkers such as Adorno, Benjamin and Lukács – referred to nature as transformed by human activities and re-naturalized (primarily as a function of ideology) as culture. Culture becomes a second order nature in, for instance, Walter Benjamin's Parisian Arcades, the 19th century predecessors of shopping malls. For Adorno, the culture industry is understood as an important element of a reified form of second nature, which individuals come to accept as a pre-formed social order. The cultural industry as second nature is a rather dark prospect of a mediated society in which the possibilities to develop a capacity to think critically about

26. Mary Ann Doane, "Technophilia: Technology, Representation, and the Feminine," in *The Gendered Cyborg: A Reader*, ed. Gill Kirkup et al. (London and New York: Routledge, 2000), 120.

one's social and political conditions are systematically obstructed. A similar notion of second nature as culture is expressed by Fredric Jameson, on the very first page of *Postmodernism*, where he suggests that "nature is gone for good", and that ""culture" has become a veritable "second nature.""[27] If, for Benjamin, there are places beyond the image culture of the Arcade where something other than culture's second nature can be found, Jameson shifts the focus by claiming that nature is gone forever and only culture remains. However, the concept "second nature," in both the tradition of Frankfurt School critical theory and the postmodern writing of Jameson, imply that there is, or used to be, a first-order nature. The loss of first nature here marks a loss of, along with a nostalgia for, a time when nature contained more of the sublime power ascribed to it by Romanticism, able to overwhelm human subjectivity, agency, and the alienating force of industrial production.[28]

Second nature has also been described as arising from a loss of immediacy, of "sensuous similarity," and involving a form of adaptive behavior by which humans use representations to adapt and assimilate themselves to their surroundings.[29] The child careening across the garden, arms rigidly outspread, imitating a jet plane; through such bodily acts of appropriation, the border between self and other – first and second nature in a secular sense – become porous, open to ongoing (re)negotiation. Such acts are (arguably) less concerned with dominating nature, than with softening the Cartesian categories of subject and object; to assert difference by,

27. Fredric Jameson, *Postmodernism, or, The Cultural Logic of Late Capitalism* (London: Verso, 1991), ix.
28. Franklin, Lury and Stacey argue that "many of the debates on "second nature" in the past have strong parallels with questions about nature and the natural in the context of globalization; indeed it is precisely certain kinds of questions about "nature" and "naturalness" which inform understandings of the global, such as the idea of global warming." They point out that many contemporary environmental critics argue that nature is no more in a globalized world in which the entire planet has been made artificial, whereas others rather convey the idea that nature is nothing but an empty sign signifying its own loss (in relation to which nature has been displaced by technology). The authors suggest that these accounts of (second) nature, as something which has followed a supposedly natural (first-order) nature, evoke a habitual reflex of nostalgia. See Franklin, Lury and Stacey, *Global Nature, Global Culture*, 21.
29. Michael Kelly, ed., "Mimesis," *The Encyclopedia of Aesthetics*, Vol 3 (Oxford: Oxford University Press, 1998), 236.

paradoxically, exploring the contours of similarity (arms make good jet wings, legs can run as fast as jet engines etc.) In other words, subjects who assimilate the objective world, rather than anthropomorphizing it in their own image, further problematize the positing of a "second" nature, with its assumption of a divisible first or third (or fifty-fifth) nature.

A similar idea is explored in Winnicott's notion of the infant's "transitional objects" that are felt to be both "me" and "not me." Here the term "potential space" mirrors that of "second nature," denoting the liminal zone between two worlds. Transitional objects such as toys or tools allow us to explore and are classified as such by infants or adults alike. This prompts Winnicott to conclude that all cultural objects and artifacts need to be understood in terms of how they first appear in play. Such a "me/not-me" learning polarity allows us to play with a new pattern or idea, testing it, staging it, distorting it, and (if it proves to work) practicing with it to internalize its pattern and add to our repertoire of skill.[30] (The philosophical dialogue and literature more generally may be said to operate in a not dissimilar manner.)

In psychosomatic transplant literature, Winnicott's notion of transitional space has been applied to the psychic integration of a transplanted organ on its donor. Years after a lung transplant, a patient has been shown to experience the received organ and donor as transitional objects. The patient perceives the donor as a living, omnipotent person (with ideal personality traits), helping the patient to cope with everyday life. The transplanted lung is thus experienced as belonging both to donor and recipient, a literal but ambiguous instance of "second" nature.[31] There is thus an ontology of grafting (or a philosophy of hybridity) to be explored further.

Our acquisition of behavior in a given context involves a process of acculturation which in turn calls into play our powers of imitation, and imitation (as we have seen in the analogy of the child pretending to be a jet plane) involves an acknowledge-

30. See Donald Woods Winnicot, *Playing and Reality* (London: Tavistock, 1971).
31. See Lutz Goetzmann, "'Is It Me or Isn't It?' Transplanted Organs and Their Donors as Transitional Objects," *The American Journal of Psychoanalysis* 64, 3 (2004): 279-289.

ment of the porousness of the membrane between various first and second order natures. Our experience of space, for example, is culturally determined, yet our everyday behavior, which takes place in actual spatial conditions, convinces us that our values and behavior, unconsciously as these may be mainly experienced, are natural. Summers comments:

> To be sure, these certainties are grounded in the conditions of our physical existences and in the limits and possibilities of these conditions; but they must just as surely be grounded in the "second nature" we learn as members of one culture rather than another. Real space, in always being concrete and lived, is thus both natural *and* conventional; that is, it belongs both to nature and to *second nature*. We learn to walk, talk and gesture, but we must also learn these things as members of cultures, and so learn to do so in specific ways, to which certain values and meanings are attached.[32]

As Aristotle understood, imitation in a general sense – copying examples – is the primary means of our socialization, helping us preserve and transmit cultural practices and values and thus cultures per se. Habits – what we learn to do by *second* nature, as a matter of course, through repetition and familiarity – are learned from participating in a group or set of practices. Our behavior (i.e. our "nature") is thus a confirmation, or attempted disavowal, that we inhabit a group and maintain our membership of that group by performing there the activities that we have learned.[33] A distinction thus emerges not so much between "given nature" and "second nature" as between spatial conditions and the cultural practices that are continually shaping and distorting such conditions. In learning our "place" in this combination – spatial experience and cultural constructions – the horizons of our power and desires are defined.

32. David Summers, *Real Spaces: World Art History and the Rise of Western Modernism* (London and New York: Phaidon Press, 2003), 53.
33. Summers comments, "'To inhabit,' in short, is to live in a social space both formative of and fitted to our second nature." Summers, *Real Spaces*, 55.

Is it in the interaction between nature and culture that we come closest to a first-order nature? Or do we need another terminology altogether? Franklin, Lury and Stacey give us an instructive example of "nature seconded" in a contemporary cosmetic product called "virtual skin" – a transparent skin foundation marketed as "skin in a bottle," which acts and feels completely natural: "As transparencies, nature and culture mimic each other's qualities such that they can hardly be differentiated, while the difference between them is precisely what makes this seconding or substitution desirable. Instead of covering the other up, culture acts and feels like nature."[34] If cosmetics are regularly thought of as assisting nature, of giving nature a helping hand as it were, "virtual skin" seconds nature further by letting culture take the place of nature, masquerade as nature, and seemingly erase the line between human skin and artifice.

The contributors to *Second Nature* are similarly investigating transfers, grafts, hybrid reformulations, or surreptitious cross-traffic between borders such as "nature" and "culture," "real" and "artifice" – that is to say, the ways in which such categories are formed and de-formed, stabilized and destabilized. As such, *Second Nature* foregrounds a view of nature and culture as something which reciprocally shape one another, letting the term "second" deliberately mark that which was never natural, the impossibility of a nature of the first order. Not only does culture shape the way we understand and experience nature, or become second nature in a naturalized form (as suggested by, for example, the Frankfurt School and Jameson), but, more importantly, if nature and culture form each other, nature itself becomes second nature, and the nostalgia associated with loss of (first) nature dissolves into thin air.

The notion of "nature" as something that was always invented, produced, and reproduced (rather than discovered) is not new, but rather a central constituent in, for example, feminist science and technology studies.[35] In decoupling nature from

34. Franklin, Lury and Stacey, *Global Nature, Global Culture*, 25.
35. See, for example, Haraway, *Simians, Cyborgs and Women*; Evelyn Fox Keller, *Reflections on Gender and Science* (New Haven; CT and London: Yale University Press, 1985); Evelyn Fox Keller, "The Gender/Sci-

the natural, another tight coupling is undone, namely that between woman and nature, or, woman-as-nature. If there was never a first nature, a nature all natural, women were never *of* nature or *closer to* nature than men. Nonetheless, the discussion of woman *as* nature and somehow more definitively embodied is a persistent theme throughout the history of philosophy, science, and medicine.[36] In her reading of the famous late nineteenth-century statue in the Parisian medical faculty "Nature Unveiling Before Science," Ludmilla Jordanova argues that science and medicine were based on the (masculine) scientific unveiling of (female) nature: "Woman, as the personification of nature, was the appropriate corpse for anatomy, which was not just literally male in that its exponents were men, but was symbolically male in that science was also the masculine practice of looking, analyzing and interpreting."[37] Woman as nature, through the many overlays between nature and (female) sexuality/reproductive capabilities, simultaneously points at a masculine mastering and manipulation of "her," as well as an un-mediated, wild and dangerous habitat yet to be penetrated and tamed. The recognition of how the liaison between woman and nature is in need of continuous creation simultaneously reveals the connection between masculinity and scientific machinery as equally arbitrary. Once disengaged from the supposedly masculine script of technology, alternative couplings between women and technologies become possible.

ence System, or: Is Science to Gender as Nature is to Science?," in *Feminism and Science*, ed. Nancy Tuana (Bloomington; IN: Indiana University Press, 1992); Carolyn Merchant, *The Death of Nature: Ecology and the Scientific Revolution* (Berkeley; CA: Harper and Row, 1980); Nelly Oudshoorn, *Beyond the Natural Body: An Archeology of Sex Hormones* (London and New York: Routledge, 1994); Londa Schiebinger, *Nature's Body: Gender in the Making of Modern Science* (Boston; MA: Beacon Press, 1993).

36. See, for example, Donna Haraway on the notion of "nature" as invented, not discovered, and as such lacking every "natural" alliance with women and between women: There is nothing about being "female" that naturally binds women. There is not even such a state of "being" female itself, itself a highly complex category constructed in contested sexually specific discourses and other social practices." in Haraway, *Simians, Cyborgs and Women*, 155.

37. Ludmilla Jordanova, *Sexual Visions: Images of Gender in Science and Medicine Between the Eighteenth and Twentieth Centuries* (New York: Harvester Wheatsheaf, 1989), 110.

Nature, much like culture, is something that is shaped and re-shaped. Rather than settling for a radical constructivist approach to nature, according to which (cultural) technologies make and mark the (natural) world in ways that fully reduce nature to culture, this book examines the material specificities of nature in ways that suggest the limits of constructivism. Following feminist physicist Karen Barad's posthumanist-materialist account of performativity and the ways in which nature comes to matter (or materializes) in scientific practices, nature is approached neither as raw material nor mere effect of cultural practices:

> Nature is neither a passive surface awaiting the mark of culture nor the end product of cultural performances. The belief that nature is mute and immutable and that all prospects for significance and change reside in culture is a re-inscription of the nature/culture dualism that feminists have actively contested. [...] Feminist science studies scholars in particular have emphasized that foundational inscriptions of the nature/culture dualism foreclose the understanding of how "nature" and "culture" are formed, an understanding that is crucial to both feminist and scientific analyses. They have also emphasized the notion of "formation" in no way denies the material reality of either "nature" or "culture."[38]

If nature is understood not as passive pre-existing matter on which culture acts as an active inscription, there is a need to rethink nature in ways that allow for the materiality of nature to play an active role in its own becoming. There are similar balancing acts within the ongoing discussion in feminist theory on ways of "thinking the body" as a border phenomenon of nature and culture – and of sex and gender, as it were. Undoubtedly, gender has been useful in contemporary feminist thought – pointing out, for example, how "masculinity" and "femininity" are cultural and historical formations (that, logically, might be subject to change).

38. Karen Barad, "Posthumanist Performativity: Toward an Understanding of How Matter Comes to Matter," *Signs* 28, 3 (2003): 827-828.

Judith Butler moved the discussion of gender several steps on by understanding gender as not only the cultural signifier of sex, but also the performative machinery by which the illusion of sex as a pre-discursive given is maintained (paralleling the radical constructivist approach in relation to which nature ultimately is an effect of cultural performances). Along the lines of a Foucauldian register of power, bodies in Butler's view are materialized through a series of reiterated language acts in ways that makes corporeal materiality an effect of power – but hardly an active force of processes of materialization.[39]

Alongside Butler's insights, it is important to recognize that the body is not merely a blank, passive surface open to societal intextuation (social, juridical, medical, disciplinary); its specific modes of materiality (none of which are innocent) are significant. Far from suggesting that bodies are pre-given objects of nature, and that there would be a "real" body on the one hand and its cultural re-presentations on the other, feminist philosopher Elizabeth Grosz conceptualizes the body as *an active open materiality* and a site for political, social, and cultural struggle. As a way of thinking about materiality as a principle of differentiation, as something that makes a difference for how bodies become meaningful, Grosz draws a parallel between corporeal matter and that of writing practices:

> The kind of model I have in mind here is not simply then a model of an imposition of inscription on a blank slate, a page with no "texture" and no resistance of its own. As any calligrapher knows, the kind of texts produced depends not only on the message to be inscribed [...] but also on the quality and distinctiveness of the paper written upon.[40]

39. There are instances in Butler's theorization of materiality in which the body "while in language, is never fully of language," see Judith Butler, *Bodies that Matter: On the discursive Limits of 'Sex'* (New York and London: Routledge, 1993), 67. As the subtitle "On the discursive Limits of 'Sex'" suggests, this argument opens up for ways of thinking bodies that although never being able to escape signification, yet inhabit a domain of "radical alterity," which, in a paradoxical sense, does escape signification.
40. Elizabeth Grosz, *Volatile Bodies: Toward a Corporeal Feminism* (Bloomington and Indianapolis: Indiana University Press, 1994), 91.

Different technologies of "inscription," to echo Foucault, certainly constitute bodies as culturally and historically specific. Then again, in an attempt to challenge the boundaries of constructivism, an exploration of the technological and cultural shaping of bodies is not enough. It is important also to examine how sexually specific yet hybrid bodies come to matter, and will matter differently, partly due to their very materiality. In her later work on Bergson and Darwin, Grosz relocates her argument, from its previous investment in human bodies and/as culture, to intersections of nature and culture, life science and cultural theory.[41] This reorientation, then, configures nature, not as that which limits culture, but rather as that which makes it possible.[42] It is a relocation of the argument that also potentially expands the discussion of corporeality beyond the bodies of humans, to also include non-human bodies and matter. In a similar way, Barad argues that feminist philosophy of the body (in her case primarily Butler's notion of performativity) can be productively de-anthropomorphized and put to use for science studies. Such an expansion of frameworks originally created to account for human subjectivity and embodiment not only makes possible investigations of the materiality of nature in non-essentialist modes, but also allows for innovative ways of thinking nature as border phenomena.

What happens, then, to the traditional coupling of woman and nature at a time when nature is produced, increasingly, by means of biotechnologies and computer technologies? If nature is always already second, yet consisting of an open and active materiality, where does this leave bio-politics and the ways in which specific bodies are made and mark(et)ed? What is the status of nature and the natural in a post-human framework in which bodies are technologically invaded and extended to the point where

41. Grosz, *The Nick of Time.*
42. In her work in the interstices of nature and culture, Grosz continues to investigate the limitations and limits of constructivism and argues: "We need to return to, or perhaps invent anew, the concepts of nature, matter, and life, the most elementary concerns of the cosmological and the ontological, if we want to develop alternative models to those inscriptive and constructivist discourses that currently dominate the humanities and the social sciences, in which the transformation of representation is the only serious political issue." Grosz, *The Nick of Time*, 2-3.

it becomes meaningless to think of the corporeal as somehow bounded by the skin? Judging from recent heated debates around technologies of (bio)reproduction, such as IVF (*In Vitro Fertilization*) practices, biological material – no matter how dislodged from its "body of origin" – is understood and regulated based on a range of cultural assumptions about what is "natural" and ethically defendable.[43] *Second Nature* seeks to explore the ways in which nature is produced and reproduced by technology, with particular attention to the many parallels and intersections between digital reproduction and human reproduction, hitherto largely neglected, as far as we are aware, in discussions of reproductive technologies. Our aim is that the anthology should also investigate our continuing attraction to second nature/s in the guise of the copy, the virus, the sample, and the clone. Science and technology studies and new media studies each address the question of how technologies of reproduction alter the meaning of concepts such as *origin*, *original*, and *originality* and how the borders between what we think of as "authentic" and "fake," "natural" and "artificial," are under constant negotiation and transformation. Discussions of "life" on one hand, and "information" on the other, converge in the idea of a society reducible to "code" – whether this means the code of life (DNA) or the code of information (computer code). We might also reflect on the significance of the fact that (post)modern biology is based on computer technology. Biological models are tested in virtual environments, and digital systems have been indispensable for the decoding of the letter combinations of DNA molecules (G, T, C, A). In the world of medical science, bodily materials such as blood and cells blends with fiber optics and the sequencing processes of computer technology.

43. See, for example, Mette Bryld, "The Infertility Clinic and the Birth of the Lesbian: The Political Debate on Assisted Reproduction in Denmark," *European Journal of Women's Studies* 8, 3 (2001); Sarah Franklin, *Embodied Progress: A Cultural Account of Assisted Conception* (London: Routledge, 1997); Sarah Franklin and Helene Ragoné, ed., *Reproducing Reproduction: Kinship, Power, and Technological Innovation* (Philadelphia: University of Pennsylvania Press, 1998); Nina Lykke, "Are Cyborgs Queer? Biological Determinism and Feminist Theory in the Age of New Reproductive Technologies and Reprogenetics," (paper presented at 4th European Feminist Research Conference, *Body, Gender, Subjectivity: Crossing Borders of Disciplines and Institutions*. Bologna, September 28-October 1, 2000).

In research on artificial life (A-Life), there are even more intricate fusions between info tech and biotech. If research in artificial intelligence (AI) has strived to simulate intelligence in a controlled, logical, top-down manner, then A-life is rather a bottom-up enterprise about growing and teaching organisms to develop and adapt on their own. Simulation and reproduction of various life forms is a growing trend in areas such as digital literature, new media (bio)art, and bio-computing. In spite of these interesting convergences, little has been said and done in the discursive spaces between digital- and life-creating reproductive technologies.[44] In *Remediation*, Bolter and Grusin (2000) argue for creative possibilities in contemporary digital media convergences, in which previously separate media forms and genres mix and create hybrids. With technologies of reproduction increasingly intervening in both works of art and (works of) bodies, how does this affect our understanding of art, life, and artificial life? What parallels might we draw between the (artificial, transplanted, once removed) nature of "life itself" and the emergence of new forms?[45]

Mid-way through our work on this book, an exhibition opened at the UCLA Department of Design Media Arts, bringing together works that in different ways reflect on the multiple connections between nature, technology, and mediation; be it simulations of nature, the use of artificial life principles to create pro-

44. For exceptions, see for example Anneke Smelik and Nina Lykke, ed., *Bits of Life: Feminism and the New Cultures of Media and Technoscience* (Seattle, WA: Washington University Press, 2008); Eugene Thacker, *Biomedia* (Minneapolis, Minnesota: University of Minnesota Press, 2004).
45. See Sarah Franklin for a discussion of "life itself," drawing on Georges Canguilhem's cultural analysis of life as a productive force, as well as Michel Foucault's notion of biopower. Franklin concludes that "we arrive at a simple sequence: nature becomes biology becomes genetics, through which life itself becomes reprogrammable information." In Franklin, Lury and Stacey, *Global Nature, Global Culture*, 190. Franklin argues that the metamorphosis of life itself has stripped nature of its function of foundational limit or force in itself: "Nature, we might say, has been de-traditionalised. It has been antiquated, displaced and superseded, and now it is only a trope – a mere shadow of the referent it used to be. That does not mean it is less useful, as we have already argued, but nature is in spin. In the place formerly occupied by "natural facts" is a new frame of reference, an offspring in the genomic era, which is life itself – now orphaned from natural history but full of dazzling promise." In Franklin, Lury and Stacey, *Global Nature, Global Culture*, 190-191.

grammed nature, or the use of optics to show how perceptions of nature are mediated (by light and by the eye). The exhibition was called *Second natures*. In one of the essays accompanying the event, the curator Christiane Paul notes: "While we casually continue to use the word nature to refer to the "outdoors" as a living habitat, it is highly debatable whether nature, in the sense of a realm untouched and uninfluenced by civilization or artificiality, still exists. Every aspect of our environment has been profoundly affected by centuries of civilization and our use (or exploitation) of natural resources. Nature has become "processed" or even designed."[46] Nature by design – a second nature of sorts – refers to technological processes of visualization, such as the weather forecast or satellite photography, by which nature is mediated and simulated as "data." This new form of nature, Paul argues, is becoming exceedingly naturalized, habitual, and familiar. It is becoming second nature. (Paul, 2006). Or, more precisely, second nature is becoming (same as it ever was) second nature.

The exhibition *Second natures* and the collections of essays in this book, *Second Nature*, share certain sensibilities and ways of understanding and exploring the connectedness and intimacy between "natural" and technological processes. But instead of asking the question of whether nature as untouched by civilization and the artificial still exists, this collection of essays rather poses the question: did it ever? Our point of departure is an understanding of nature and the natural as always already artificial, yet neither immaterial nor lacking a force of its own. Instead of viewing nature as something "out there" for (wo)men to discover, the raw materials for a nature by design, this book departs from the assumption that there was never a way of viewing or knowing nature beyond various technologies and mediations. Put differently, first nature only exists through an interaction with second nature. According to this understanding, nature is always second; constructed, created, made. We create William Wordsworth at the moment we peer through his lens on the Lake District in our imagination. The issue at stake is not whether nature and culture, or, more aptly

46. See http://projects.design.ucla.edu/exhibitions/SecondNatures/cp.html (accessed June 21, 2010).

nature and technology, are becoming all the more entwined based on the assumption that they were previously, at some point, apart. Instead we might ask what kind of technologies have produced nature – and how have such technologies defined the kind of nature being produced. Further, we might ask why and how nature has come to matter once again, as both material and agency in artistic and scientific practices, and how might we rethink our various natures accordingly.

Contributions to *Second Nature*

JOHN MONK traces, from the sixteenth century to the present, different attempts to account for the liveliness of animals and people through analogies and analogues, attempts that typically synthesize the prevailing physical, biological, and technological discourse of the day and are concerned less with what things are made of than how they collect, process and exploit information. With the computer and communication networks – sign systems *par excellence* – organizations became information processors and living beings became sign systems constructed from the data provided by sensors in shops, hospitals, schools and government offices. Monk highlights the risk of becoming so steeped in analogues and analogies, and seduced by their simplicity, that we neglect in our performative symbolic actions the present and the local – and the pains, anxieties, morality and mortality that loom outside of the analogue.

Drawing on her residency at the SymbioticA Art and Science Collaborative Research Laboratory in the School of Anatomy and Human Biology at the University of Western Australia, BOO CHAPPLE dissects the politics of control that come into play when art meets the biotech industry. As the biotech industry seeks to commodify and manipulate life in new ways, so the artwork becomes less concerned with creating a picture or a world, than with "intersecting with and framing processes and situations that already exist" resulting in "a constant reciprocality between biological (re)production and artistic production."

RON BROGLIO's chapter discusses cattle as simultaneously part of nature and of culture – "hinge objects that work between the realms of the artificial and the natural, with each jostling in relation to the other." Tracing the complex slippage between nature and artifice in modern cattle breeding, he examines how functional genomics has converted the outer surface of cattle hair and skin to genetic information that yields a look "inside" the animal, introducing in the process complex legal and social questions that highlight the continuum of technological enframing of the animal body.

MARIA CHATZICHRISTODOULOU explores notions of *presence*, with its assumption that the corporeal body is an "original" whereas all its representations, reproductions and/or extensions are "artificial," an *absence* in networked encounters to describe the condition and experience of situating (for presence) or excluding (for absence) one's corporeal body and "aura" within a specific spatio-temporal context, and a set of relationalities that include others. In questioning this presence/absence dichotomy, the author draws on literary theorist Katherine Hayles' proposed alternative or complementary dialectic based on notions of pattern and randomness, illustrating the implications of this shift through a discussion of Entropy8Zuper!'s award-winning performance/software/net art piece *Wirefire* (www.entropy8zuper.org/wirefire/).

With new technologies of reproduction rapidly transforming cultural understandings of kinship, family, body, sexuality, and origin(s), JENNY SUNDÉN'S essay critically explores medical simulations of birth, the lifelike, and techno-corporeality through a close encounter with a blonde, white, birthing machine: S565 NOELLE™. An investigation of techno-corporeal becomings in medical simulation, Sundén provides an account of the "birth" of birth simulators themselves, investigating the status of the "real" as well as of "realism" in the simulation world; she explores historical parallels or predecessors to medical simulators in the shape of anatomical wax models, and, in response to recent feminist discussions of difference in terms of "intersectionality," explains how this particular reproductive machinery is entwined with issues not merely of sexual difference and sexuality, but also of race and national belonging.

KARIN WAGNER'S contribution explores net art, parody, and genetics through Virgil Wong's *GenoChoice*, a web site that parodies current medical discourse around patient choice to question the social, cultural, and political dimensions of biotechnology and reproduction. Wong's art work encourages us to consider to what extent human identity can be represented as pure (genetic) information and how might cultural significations of kinship and family be altered in a society of code. FRANÇOIS-JOSEPH LAPOINTE and MARTINE ÉPOQUE investigate the implications of *in silico* as well as *in vitro* creative processes for dance composition, a form of trans-specific generative art that not only redefines the role and function of the artist (or meta-artist), but also raises important questions about authorship and collaboration. Focusing on their own "post human" choreography, their chapters present an application of trans-specific collaboration in dance.

TIMOTHY WEAVER examines the impact of applied biomimetics (including bioprospecting, biophilia, biopiracy and emulated biosemiotics) across a spectrum of creative and technological-mediated processes. Using three comparative lenses – the scientific characterization of natural biological systems; historically-rooted sociocultural performances/practices and contemporary engineered-design and creative media-based expression – his paper seeks to align the biomimetic approach to the appropriation and emulation of natural systems within traditional and post-traditional ecologies as a means of characterizing the distance between origin and derivation in this creative practice.

ROLF HUGHES examines the practices of "bioartists," "transdisciplinary" theorists, "generative" or "evolutionary" designers and others, and the challenge such work poses for theoretical, critical and aesthetic judgements founded on mono-disciplinarity with its deep-seated, philosophically and culturally privileged concepts of human agency and responsibility (including authorship, imagination, authenticity, originality, and accountability). In place of such disciplinary and philosophical preoccupations with origins, his paper explores the potential metaphors, disciplinary hybrids, and theoretical practices that arise from a conception of cultural production based on transdisciplinary collaboration and "posthuman" creativity.

Finally, EUGENE THACKER's essay explores the relation between animality and biotechnology, focusing both on contemporary issues, such as "biodefense," as well as historical issues, such as the Medieval bestiary. Animality – as the human capacity to "think the animal" – is found to exist within the networks and passages that both constitute and threaten social, economic, and political life. Thacker presents three cultural relationships between human and animal, relationships that not only challenge us to rethink the animal, but also the human, and asks whether the human-animal relationship occupies an indeterminate zone between the everyday and the exceptional, the ordinary and the extraordinary.

Bodies, Machines, Politics

John Monk

In the sixteenth century liveliness was attributed to animal spirits in a notion borrowed from the Greeks.[1] Moreover to explain how the animal spirits caused bodily movements, writers commonly drew parallels between machines and bodies. Often they also compared machines and bodies with forms of political organisation. It was as though people saw images of themselves, and parts of themselves, mirrored as machinery or bureaucratic apparatus.[2]

A little later, Descartes aimed to describe what characterised life. He took life to be action and the nervous system to be the bearer of activity, and he illustrated his standpoint with analogies based on machines of his day. Today life is linked with genetics and our analogues have subtly shifted from mechanical material handling to the contemporary signal handling machinery of computer and communication technology.

The goal of analogies is to hint at points of contact between two domains. An analogy has a characteristic of symbols in that one thing stands in place of another. Deploying an analogy or analogue takes the mode of expression from one domain and appends it to another. An analogy or an analogue affords a different view; it reflects, distorts and reveals; it transposes vocabulary and grammar and a repertoire of effects and relationships into an unfamiliar, unexpected or previously unimaginable setting; it transplants context, rhetoric and logic. It subtly draws in connotations, underlines essential features and shows what is inessential by interchanging parameters such as persistence and transience, time and space, social and individual, natural and artefactual.

Analogies are components of a variety of language games which interrupt the flow of logic in an explanation or argument.

1. Max R. Bennett, "The Early History of the Synapse: From Plato to Sherrington," *Brain Research Bulletin* 50 (1999): 95–118.
2. Anon, "Organic and Mechanical Metaphors in Late Eighteenth-Century American Political Thought," *Harvard Law Review* 110 (1997): 1832–1849.

They may simply provide a diversion, enliven the prose or even conceal a logical weakness. But an analogy can also augment a discourse with a fragment of reasoning or a prototype of an axiom or relationship from another literary bailiwick and thus generate novel results and discoveries, which through customary use can become an intrinsic, literal part of their adopted domain.

To provide a tenable contextual shift, an analogy needs firm roots derived, for instance, from an enduring fable, a revered text or common experience. In the sixteenth century, European intellectuals were inclined to accept that "human conduct is 'right' only if . . . it is *natural* or *according to nature*."[3] References to Nature provided and continued to provide a safe and sure anchor for analogies, so much so that writing in 1686, Robert Boyle, complained about the word's overuse. "Nature", he remarked, is "ambiguous, and so often abus'd, a term."[4] In various ways Nature was coupled to the body and to machinery and a knot of analogies emerged linking machines, bodies and social structures. This collection of analogies provided a heritage that persists in the fields of technology, politics and the biochemistry of bodies and which continually adjusts to accommodate scientific, political and technological developments.

Bodies and politics

In *Measure for Measure*, first performed in 1604, Shakespeare hinted at an anatomical analogy when Lucio spoke of the "nerves of the state" to refer to the politics and business of an administration.[5] Earlier, William of Conches, the author of the twelfth century *Philosophia,* introduced an "analogy between the excellences of the human body and the body politic". In this analogy, the head was the citadel, "the arms represent soldiers guarding the whole, the stomach and knees represent artisans and workers, the bones

3. Stephen Toulmin, *Cosmopolis* (Chicago: University of Chicago Press, 1990), 68.
4. Robert Boyle, *A free enquiry into the vulgarly receiv'd notion of nature* (London: John Taylor, 1686), 389.
5. William Shakespeare, *Measure for Measure*, in *Mr. William Shakespeare's Comedies, Histories, & Tragedies* (London: Isaac Laggard, and Ed. Blount *1623*), Act 1, Scene 4.

and blood, business men and the feet, farmers."[6] Hobbes exploited a similar analogy when, in the opening of his Leviathan published in 1651, he compared the state with an "Artificiall Man". The state, Hobbes proposed, derives its life and motion from sovereignty, which, he maintained, is analogous to an "artificial soul". Hobbes then offered additional likenesses; for example, "the magistrates and other officers of judicature and execution" were equated with bodily "joynts"; reward and punishment were likened to the action of the nerves, and so on.[7]

These descriptions were bolstered by an association between the body, nature and a deity, a doctrine neatly encapsulated in the title page of a book published in 1665 when it promised "a new and true description of the law of God (called nature) in the body of man."[8] By linking the components of the state to the components of the body, and reinforced by his view that man is "that rational and most excellent work of Nature", Hobbes asserted the naturalness of his political arrangement, and "Nature", as has often been the case, was exploited to bestow authority on an analysis.

Rousseau, in his *Discourse on Political Economy* published in 1755, also coupled the body and the state and suggested "the laws and customs are the brain" and, with the hint of an invitation to exploit the analogy further, the brain is, he noted, "the source of the nerves and seat of the understanding, will and senses."[9] Much later, in the last quarter of the nineteenth century Herbert Spencer introduced an elaborate analogy of an organism to describe social relationships and processes. He discussed, in some detail, the analogy between elements of social organisation and the nervous system, not only its structure but, borrowing from developments in theories of biological evolution, he reasoned that

6. Elspeth Whitney, "Paradise Restored. The Mechanical Arts from Antiquity through the Thirteenth Century," *Transactions of the American Philosophical Society, New Series* 80 (1990): 1–169.
7. Thomas Hobbes, *Leviathan, or, The matter, forme, & power of a common wealth ecclesiasticall and civil* (London: Andrew Crooke, 1651), 1.
8. William Drage, *A physical nosonomy* (London: R. Tomlins & Geo. Calvert, 1665).
9. Jean-Jacques Rousseau, *A Discourse on Political Economy*, trans. G. D. H. Cole (New York: E.P. Dutton and Co., 1950), 252. Original published in 1755.

> the nervo-muscular apparatus which carries on conflict with environing organisms ... is developed by that conflict; so the governmental-military organization of society... evolves along with the warfare between societies.[10]

Both parts of the analogy are speculative but somehow they gain support from one another and dampen challenges to Spencer's assertions. The analogy here is more a distraction than a justification, but it does intimate that, for Spencer, there was naturalness in his explanation of social evolution and that continual conflict and consequent social change was inevitable.

Machines and bodies

Hobbes, as well as referring to governance, also equated bodily parts with mechanical components by treating the heart as a spring, the nerves as strings and the "joynts" as wheels, so through the relationship with machine parts he conjured up an image of a well-ordered anatomy for both the body and the state. Like the body, machines gained positive moral and religious associations.

A fourth century author drew an analogy between steps taken in manufacturing and the phases of the creation of the Universe[11] and also mentioned the admiration that an artificer might receive as a consequence of the "industrious intelligence" displayed in an artefact. However, in the sixth century, Augustine, who had inherited from antiquity a disdain of technology, reckoned that, for most people, knowledge of the subject was only worthwhile for interpreting "figurative locutions based upon them."[12] A change is evident in the twelfth century when Hugh of St. Victor founded a school which admitted the mechanical arts as a major branch of knowledge. Gradually, with the wider circulation of printed texts, Aristotelian and Arabic influences gained ground. Together they

10. Herbert Spencer, *Principles of Sociology*, ed. Stanislav Andreski (London: Macmillan, 1969), 85. Original published in three volumes between 1876 and 1896.
11. St. Basil of Caesarea, "Hexameron" in *Nicene and Post-Nicene Fathers, Series II, Vol. VIII*, ed. Philip Schaff and Henry Wace (Peabody, Mass: Hendrickson Publishers, 1994).
12. Augustine, *De Doctrina Christiana*, trans. D. W. Robertson (Indianapolis: Bobbs-Merrill, 1958), bk. II, ch. 30, 66.

raised the spiritual values attributed to machines. Hugh's followers, for instance, regarded the practice of arts as a route to salvation, which remedied man's physical weaknesses resulting from the Fall of Adam and Eve. In a more secular vein, technology was regaled as an adjunct to the noble pursuit of science or as an activity that required the practitioner to engage in the valued activities of deliberation and reasoning and "only later manifest themselves in a work". And in a translation of Francis Bacon's words, Cressy Dymock when promoting his machine and referring to mechanical inventions wrote, in 1651, "The Introduction of Noble Inventions seemeth to be the very chief of all humane actions" and, he added, "new inventions are as it were new creations, and imitations of Gods own works".[13] The mechanical Arts thus grew in stature and regained respectability as a source of analogy though some ambivalence remained.[14]

In his *Rules for the Direction of the Mind*, René Descartes (1596–1650) gave an explicit justification for the use of analogy. When we are trying to explain something that is outside our experience, like the inaccessible internal workings of the body, rational argument, he claimed, is unproductive.[15] Instead of an appeal to reason, Descartes called for the kind of imaginative abstraction that leads to a comparison with a familiar setting. He contrasts the artist, who needs to practice a particular skill with the scientist who can exploit knowledge gained about one topic to provide hints about what there is to know about another. In Rule One he explained that in science "the knowledge of one truth does not ... hinder us from discovering another; on the contrary it helps us".[16]

This precept supported Descartes' exploitation of a thorough understanding of machines when he described the mysterious workings of the body and openly described the "mechanism"[17] of

13. Cressy Dymock, A Passage out of the Lord Verulan's Novum Organon in *An invention of engines of motion...* (London: Printed for Richard Woodnoth, 1651). Original in Latin and published in 1620 art.129. Verulan (or Verulam) is the title bestowed on Francis Bacon.
14. Whitney, "Paradise Restored."
15. René Descartes, "Rules for the Direction of the Mind," in *The Philosophical Writings of Descartes Vol.1*, trans. John Cottingham, Robert Stoothoff and Dugald Murdoch (Cambridge: CUP, 1985), 57, Rule 14.
16. Descartes, "Rules for the Direction of the Mind," 9.
17. Descartes, "Passions of the Soul," in *The Philosophical Writings of Des-*

the human body as "nothing but a statue or machine"[18] – though he did append to the body a soul that "understands, wills, imagines, remembers and has sensory perceptions."[19] Descartes' *Description of the Human Body* provided an outline of what he called "the entire machine" including a description of the nervous system. Briefly, his theory was that

> the parts of the blood that are most agitated and lively are carried to the brain by the arteries ... [and] make up a kind of air or very fine wind which is called the 'animal spirits' [that] flow from the brain through the nerves to all the muscles ... [and] inflate the muscles in various ways and thus impart motion to all parts of the body.[20]

By relating the body and machines, Descartes presumed people who are familiar with "automatons, or moving machines" would find, for instance, his explanation of the reflexes plausible.[21]

Machines, unlike bodies or organisations, are expected to be made of assemblies of components taken from a limited inventory of idealised forms – springs, levers, gears and so on – with each component having a distinguished and unwavering, definable and bounded behaviour. By drawing a parallel with descriptions of machines, Descartes illustrated how explanations of organisms and organisations might be orchestrated and simplified. To drive home a description of, for example, the mysteries of the nervous system Descartes conjured up an image of a once well-known hydraulic system, which was incorporated in "the grottos and fountains in the royal gardens", and he drew explicit parallels comparing "the nerves ... with the pipes in the works of these fountains, ... animal spirits with the water that drives them, the heart with the source of the water, and the cavities in the brain with the storage tanks".[22]

cartes Vol.1, 330.

18. Descartes, "Treatise on Man," in *The Philosophical Writings of Descartes Vol.1*, 99.
19. Descartes, "Description of the Human Body and of all its functions," in *The Philosophical Writings of Descartes Vol.1*, 314.
20. Ibid., 316.
21. Descartes, "Discourse on the Method, Part 5," in *The Philosophical Writings of Descartes Vol.1*, 139.
22. Descartes, "Treatise on Man," 100.

The analogy of the fountains not only helped Descartes' readers to grasp his view of the role of the nerves, but it also substantiated the form of the description since it portrayed functions and connections by referring to a construction on a comprehensible scale that could be described and also visited, consciously explored and experienced. Of course, few would have dreamed of conducting such a practical validation, but its possibility made the analogy more secure.

Descartes argument was shaped by a manufactured artefact that had a behaviour. He was therefore able to take his analogy beyond the constructional similarities and compare examples of behaviour in both domains. He supposed the stimulation of the senses caused bodily reactions in the same way as visitors to the Royal fountains unwittingly triggered movements, for instance by "stepping on certain tiles which are so arranged that if, for example, they approach Diana who is bathing they will cause her to hide in the reeds".[23]

He interspersed his texts with further analogies. For autonomous activities like breathing, which he believed relied on the flow of animal spirits, Descartes exploited the analogy of "the movements of a clock or mill".[24] Elsewhere he compared the natural flow of spirits in a bodily reflex with the way "the movement of a watch is produced merely by the strength of its spring and the configuration of its wheels."[25] To explain the reaction of a human foot that comes too close to a fire, he introduced a tiny theoretical fibre that tugs at the brain "just as when you pull one end of a string, you cause a bell hanging on the other end to ring"[26] and he likened the nervous reaction to touched objects to the movement of air in an organ triggered by the action of the organist's fingers.[27]

Lest his analogies led to the belief that there is no distinction between people and machines, Descartes set criteria for differentiating machines from people – one criterion implied that a machine could not "produce different arrangements of words … so as to give appropriately meaningful answers to whatever is said

23. Descartes, "Treatise on Man," 101.
24. Ibid., 100–101.
25. Descartes, "The Passions of the Soul," 335.
26. Descartes, "Treatise on Man," 101.
27. Ibid., 104.

in its presence".[28] This practical test shows remarkable similarities to the test devised by Turing three hundred years later to assess whether or not machines can think.[29]

Many authors around Descartes' time trod a line between analogy and literalism by referring to mystical animal spirits that have realistic fluid like properties. For example, in 1674 in a treatise, Malebranche frequently referred to the body as a machine which achieved controlled movement by exploiting reservoirs of animal spirits in the brain.[30] Though, in describing the actions of the muscles he proposed an alternative analogy incorporating a football that could be inflated easily yet raise heavy weights.[31]

Animal spirits were often mentioned in explanations of how bodies moved, not as an analogy of anything, but as an existent but unobservable substance. Analogy provided a link with observable substances, or at least their readily observable or familiar effects. Spirits in the body in these accounts behaved like fluids – water in the fountain, mill or clock, or air in the church organ or football – or provided a threadlike connection. And thus the body could be seen as being analogous to, for example, hydraulically or mechanically operated machinery.

A picture emerged in which the body, exemplifying Nature, is subject to the will and controlled by mechanistic reflexes and a centralised brain; furthermore, because of their similarity to machines, bodies and their behaviour were seen to be amenable to rational explanation. Additional steps connecting politics with bodies and machines through analogy thus provided the rhetoric to promote political systems operated through centrally controlled, rational and obedient administrations.

As Toulmin explains, Descartes was writing at a time that Europe was in the throes of the turbulent Thirty Years' War, which was partially the result of religious differences, and intellectuals,

28. Descartes, "Discourse on Method," 140.
29. Alan M. Turing, "Computing Machinery and Intelligence," *Mind* 59 (1950): 433–460.
30. Nicolas Malebranche, *Father Malebranche his treatise concerning the search after truth," Book 2, 2nd edition*, trans. T. Taylor (London: Thomas Bennet, T. Leigh and W. Midwinter, 1700), 52. Original treatise published in French in 1674.
31. Malebranche, *Father Malebranche his treatise concerning the search after truth*, Book 6, 82.

seeking a way out of the conflict, were struggling to create a new social order.[32] Descartes' analogies shifted the credibility of his explanations so they did not rest on religious foundations but in knowledge of working machines, or the possibility of their existence, and that knowledge was possessed by secular authorities who made judgements about which machines worked and which did not. However, Descartes' analogies referred to technologies that some writers thought met unworthy goals so Descartes, an ardent Catholic, troubled by the news that Galileo had been punished for his writings and, most likely, not wanting to offend the church,[33] did not publish his anatomical work; it was published after his death.[34]

Hypothesis

Liveliness and hence life, in seventeenth and eighteenth century Europe, had become linked to an ineffable fluid that constituted the soul. Both Locke[35] and Hume,[36] for instance, subscribed to the notion of animal spirits and their role in transmitting sensations. Much earlier in his *Lives of the Philosophers*, written around the third century, Diogenes Laërtius reported that, according to Aristotle and Hippias, Thales of Miletus "attributed souls also to lifeless things, forming his conjecture from the nature of the magnet, and of amber."[37] When rubbed, for example with fur, amber accumulates an electric charge, attracts light objects and can with sufficient effort generate a spark. Thales thus planted the thought that the attractive force of an invisible agent was connected with liveliness. In the face of such a mystery, writers resorted to analogy to fill the perplexing gulf between observable potential causes and observable effects.

Robert Boyle, in notes published in 1675, reviewed hypotheses

32. Stephen Toulmin, *Cosmopolis* (Chicago: University of Chicago Press, 1990), 98.
33. S. Gaukroger, *Descartes: an intellectual biography* (Oxford: Clarendon, 1995), 290–292.
34. Gaukroger, *Descartes: an intellectual biography*, 221.
35. John Locke, *An essay concerning human understanding*, Book II (London: Thomas Bassett, 1690), 55.
36. David Hume, *A treatise of human nature*, Vol. 1 (London: John Noon, 1739), 368.
37. Diogenes Laërtius, "Life of Thales," in *The Lives and Opinions of Eminent Philosophers*, trans. C.D. Yonge (London: George Bell & Sons, 1895).

about electrical attraction such as the conjecture provided by the Jesuit, Cabaeus, who proposed that the attractive force of amber was due to "the steams that issue ... out of Amber, when heated by rubbing" and which makes "a small whirlwind."[38] Boyle's alternative analogy likened the electrical influence to "a drop of oyl or syrup [that] hangs from the end of a small stick."[39] Newton, at the very end of his *Principia*, published in 1687, taking a more literal line wrote about "a certain most subtle Spirit" which caused forces between "electric bodies" and transmitted sensations and commands to the muscles through the nerves. However, he concluded that there have been insufficient experiments to support any theories about "this electric and elastic spirit."[40]

By 1731, Alexander Stuart thought he could "fairly conclude, that there is a fluid in the nerves"[41] and in 1767 Browne Langrish introduced another analogy inspired by the evolving steam engine to explain muscle action in which animal spirits behave like "the steam of boiling water working in the engine to raise water by fire". Langrish however rejects this as a proper explanation and assumed there is some kind of flux through the nerves that he attributed to an "*Æthereal Medium*" that flows "as quick as Lightning". To calm doubts he drew upon an analogy and suggested, "experiments on electricity ... give us an idea of the great subtilty and velocity of the nervous fluid". Yet he did not take any steps that would directly implicate electricity in nervous conduction.[42] In a lecture by George Fordyce, given in 1787, there is a hint that "the nerves are surrounded with something like electric matter, in which motion runs from the brain to the moving parts", but ultimately Fordyce rejected this and favoured a phenomenon he called the "attraction of life."[43]

38. Robert Boyle, *Experiments and Notes about the Mechanical Origine or Production of Electricity* (London: R. Davis, 1675), 2–3.
39. Boyle, *Experiments and Notes about the Mechanical Origine or Production of Electricity*, 4.
40. Isaac Newton, *The mathematical principles of natural philosophy*, Vol. 2, trans. Andrew Motte (London: Benjamin Motte, 1729), 393.
41. Alexander Stuart, "Experiments to Prove the Existence of a Fluid in the Nerves," *Philosophical Transactions (1683–1775)*, 37(1731–1732): 327–331.
42. Browne Langrish, "The Crounean Lectures on Muscular Motion, Read before the Royal Society in the Year 1747," *Philosophical Transactions (1683–1775)* 44 (1746–747): i–ii+1–66.
43. George Fordyce, "The Croonian Lecture on Muscular Motion, Read November 22, 1787," *Philosophical Transactions of the Royal Society of*

Around this time the possibility of electrical conduction over a distance was known. For instance in a succession of experiments John Gray, in 1729, with a colleague had found a way of suspending pack thread with insulating strands of silk so the pack thread could transmit electrical effects a distance of 765 feet.[44] Later M. le Monnier in 1746 in Paris reported on a device incorporating a glass phial for storing electric "virtue" that gave "a violent concussion in both the arms to 200 men all at once … holding each other by the hand" and he announced that "electricity has in this manner been carried … near 2 miles."[45] William Watson intent on discovering the conducting properties of stretches of ground created a ciruit by channelling the electricity he generated above the ground along a wire together with connections into the earth at both ends. In August 1747 in spite of practical problems, for instance, when his wire was broken "by a Man riding through them", Watson supplied electricity at one end of his circuit and detected it at the other with "two country fellows … [who] with four … gentlemen formed a chain, the first of them taking hold of the extremity of the wire". In the course of the experiment the participants received a shock and "the countrymen could by no means be prevailed upon to try the experiment again." Watson concluded "that the electrical commotion has been perceptible …, even as far as two miles."[46]

By the middle of the eighteenth century there was a growing public understanding of the properties of electricity and their use in analogy was illustrated in the novel about the exploits of Tom Jones when Black George who, to quell his anger, took his traditional remedy – "a kind of horse-medicine". Subsequently the whole family

London 78 (1788) 22–36.

44. Stephen Gray, "A Letter to Cromwell Mortimer, M. D. Secr. R. S. Containing Several Experiments concerning Electricity," *Philosophical Transactions* (1683–1775) 37 (1732):18–44.
45. Monsieur le Monnier, "Extract of a Memoir concerning the Communication of Electricity; Read at the Public Meeting of the Royal Academy of Sciences at Paris, Nov. 12. 1746 …", *Philosophical Transactions* (1683–1775) 44 (1746 – 1747): 290–295.
46. William Watson, "A Collection of the Electrical Experiments Communicated to the Royal Society," *Philosophical Transactions* (1683–1775) 45 (1748): 49–120.

> were soon reduced to a state of perfect quiet; for the virtue of this medicine, like that of electricity, is often communicated through one person to many others, who are not touched by the instrument. To say the truth, as they both operate by friction, it may be doubted whether there is not something analogous between them.[47]

Such were the developments in knowledge that by 1769 Priestley was able to write a history of electricity in which Priestley asked, "Is the fluid on which electricity depends at all concerned with the functions of the animal body?"[48] but in the brief chapter on the topic he merely asks more questions.

Kinds of electricity or analogues?

Three disparate techniques were available for creating similar effects that were later attributed to a common agent – electricity. Initially, however, although there was a suspicion of a hidden common cause but with few opportunities for comparison the descriptions of the effects could only be treated as analogies. In 1752 Abbé Nollet wrote a letter that referred to "natural electricity in the atmosphere" as distinct from "artificial electricity … excited by friction" and the Abbé indicated that he was "more interested than any body to come at the facts, which prove a true analogy between lightning and electricity."[49]

In a letter also written in 1752, Benjamin Franklin described his famous experiments with a kite and wrote that he had collected the electric charge in a glass phial and performed experiments that give similar results to those "done by the help of a rubbed glass globe" – he concluded there is a "sameness of the electric matter with that of lightning."[50] Clearly, because of the extent of the

47. Henry Fielding, *The History of Tom Jones, a Foundling* (London: G. Routledge, 1857), Book 4, Chapter 9. Original published in 1749.
48. Joseph Priestley, *The history and present state of electricity, with original experiments* (London: J. Dodsley, J. Johnson and J. Payne and T. Cadell, 1769), 468.
49. Abbé Nollet, "Extracts of Two Letters of the Abbe Nollet, … Relating to the Extracting Electricity from the Clouds Translated from the French," *Philosophical Transactions (1683–1775)* 47 (1751–1752): 553–558.
50. Benjamin Franklin, "A Letter of Benjamin Franklin, Esq; to Mr. Peter

similarities between electrical effects Franklin was convinced the effects were the result of the same "matter" and that they were not merely analogous but identical, which is, perhaps, what Nollet meant by "true analogy".

Thus the continued and expanding use of analogy resulted in analogical accounts that were indistinguishable in most respects from literal description of the domain that the analogy mined. Gradually the analogies connecting "natural" and "artificial" electricity, coalesced, the two domains merged to become populated with identically described causes, effects, relationships and objects.

Besides electricity generated by friction or atmospheric effects, there was another candidate, "animal electricity", that created similar effects which were described by John Hunter, who studied the anatomy of electric fishes.[51] These effects were also exemplified in experiments detailed in a Letter from the Mayor of la Rochelle to the French Gazette dated 1772. Apparently, a fish, a live Torpedo, "was placed on a table. Round another table stood five persons insulated ... the five persons communicated with one another ... Mr. Walsh touched the back of the Torpedo, when the five persons felt a commotion."[52] A subsequent note from the experimenter and addressed to Benjamin Franklin explained, "the effect of the Torpedo appears to arise from a compressed elastic fluid ... in the same way... as the elastic fluid compressed in charged glass."[53] A more direct comparison between the effects of the torpedo fish and the static charge on a glass tube produced similar results when John Ingenhousz experimented on a group of unfortunate sailors. Ingenhousz recounted,

> I charged a coated jar by a glass tube, and gave a shock to some of the sailors; who all told me, they felt the same sensation as when they touched the torpedo.[54]

Collinson, F. R. S. concerning an Electrical Kite," *Philosophical Transactions (1683–1775)* 47 (1751–1752): 565–567.

51. John Hunter, "An Account of the Gymnotus Electricus," *Philosophical Transactions (1683–1775)* 65 (1775): 395–407.
52. John Walsh, "Of the Electric Property of the Torpedo. A Letter to Benjamin Franklin," *Philosophical Transactions (1683–1775)* 63 (1773–1774):461–480.
53. Walsh, "Of the Electric Property of the Torpedo ..."
54. John Ingenhousz, "Extract of a Letter from Dr. John Ingenhousz,

Around the same time John Hunter showed that the organs responsible for the shock were associated with thick bundles of nerves that he supposed "are subservient to the formation, collection, or management of the electric fluid"[55] and he appeared to accept the shock was produced by the same "fluid" as produced by friction. Thus electricity once again became strongly associated with nerves and as a result of the similarities of their effects animal, artificial and natural electricity gradually came to be treated as being literally the same. A lewd satirical poem published in 1777 with the title *The Torpedo* and subtitled *A Poem to the Electric Eel* showed that popular opinion was accepting the involvement of electricity in the liveliness, at least, in men and fish, since displayed on the title page of the pamphlet was the sentence "Electricity will soon be considered as the great vivifying principle of Nature."[56] Notably, the generic term "electricity" was not qualified by any adjective such as "animal" or "natural".

Galvani gave his name to yet another way of generating electricity that emerged from his accidental discoveries. But initially, in 1791, he reported on an unplanned coincidence in a laboratory where some assistants were experimenting with electric machines that generated electricity by friction while simultaneously others were conducting experiments on dissected frogs. One of the assistants noticed a frog's muscles contracted when he touched a nerve with a scalpel. They managed to show, what another assistant had ventured, that this only "occurred when a spark was discharged from … the electrical machine". The frog's leg jerked only when someone was holding the metal of the scalpel so they tried again using "a very long wire… to … compensate for the absence of the man" and once again "the muscles contracted when the sparks were discharged". With an iron wire "of one hundred and more ells in length … muscular contractions occurred at the emission of a spark, even though the distance from the electrical machine

F. R. S. to Sir John Pringle, Bart. P. R. S. Containing Some Experiments on the Torpedo March 27 1773," *Philosophical Transactions (1683–1775)* 65 (1775): 1–4.

55. John Hunter, "Anatomical Observations on the Torpedo," *Philosophical Transactions (1683–1775)* 63 (1773–1774) 481–489.

56. Anon., *The torpedo, a poem to the electrical eel. Addressed to Mr. John Hunter, surgeon … The fourth edition, with large additions*, London, 1777.

was very great."[57] Galvani's second series of experiments tested the reaction of the frogs' legs to distant lightning and as he anticipated "the result completely paralleled that in the experiment with artificial electricity."[58]

Further chance events led Galvani to another discovery, showing that closing a circuit comprised of a prepared nerve and muscle, and a hook and a bar made from two different metals, caused the muscle to twitch. Galvani's explanation involved animal electricity, which he distinguished from "common ordinary electricity."[59]

Although the first two sets of experiments demonstrated that electrical effects were implicated in the workings of nerves and muscles and that lightning and sparks from frictional electricity caused similar effects (and presciently provided a demonstration of a potential form of wireless communication), Alexander Volta was dismissive of Galvani's work. A commentator wrote, "Mr Volta …has shewn that the conclusions, which Mr Galvani drew … are in various respects erroneous."[60] Volta also criticised Galvani's analogy which attempted to draw a parallel between muscle action and the discharge of a Leyden jar (a simple device for storing electric charge). However he thought that the discoveries involving dissimilar metals were "completely new and marvellous" and this led Volta to develop the electrochemical battery, the Galvanic cell, which made possible new kinds of investigation into electrical phenomena and provided a significant, but a frequently overlooked, component of the emerging electrical communication systems – the battery.

At a meeting, in 1830, presided over by Mark Isambard

57. Luigi Galvani, "Part I The Effects of Artificial Electricity on Muscular Motion," in *De viribus electricitatis in motu muscolari. Commentarius 1791 published in English as Galvani's Commentary on the effects of electricity on muscular motion, with notes and critical introduction* , trans. Margaret Glover Foley (Norwalk, Conn.: Burndy Library, 1953).
58. Galvani, "Part II The Effects of Atmospheric Electricity on Muscular Motion," in *De viribus electricitatis in motu muscolari. Commentarius 1791.*
59. Galvani, "Part III The Effects of Animal Electricity on Muscular Motion," in *De viribus electricitatis in motu muscolari. Commentarius 1791.*
60. William Charles Wells, "Observations on the influence, which: includes the Muscles of Animals to contract in Mr Galvani's Experiments," *Philosophical Transactions of the Royal Society of London* 85 (1795):246–262.

Brunel,[61] Michael Faraday systematically went through the phenomena associated with the different forms of electrical generation and concluded "that electricity, what ever be its source, is perfectly identical in its nature". Nevertheless he confessed that the question was not settled.[62] Electricity known only by its effects and its methods of generation eventually ceased to be divided into different analogous categories. It was accepted that, irrespective of the way it was generated, electricity, a common agent, could demonstrate the identical, rather than analogous, effects. Analogies became identities.

Electrical Communication Technologies

In the first half of the nineteenth century, practical application, in the telegraph for instance, rather than curiosity was beginning to drive investigations into electrical phenomena and the introduction of the telegraph presented new possibilities for describing or exploiting existing knowledge of the nervous system.

It is seldom obvious as to why the analogy between the nervous system, telecommunications and societal structures should be employed. Perhaps occasionally it is to take the opportunity to link tacitly technical and social systems with three political ideals: firstly, the centralisation of decision making as exemplified by the brain; secondly, the identity of the state or organisation as an integrated body, and, thirdly, the machine-like nature of administration. Or perhaps, on occasions, the technical or organisational discourse gains stature from implicit references to Nature and naturalness. The analogies also create a relationship between forms of political, biological and technological expression, and any anatomical discoveries, political events or developments of machines absorbed into accounts of related fields bestow an air of modernity and progress.

Whatever the motivation, the popular acceptance of the analogy between the telegraph and the senses is evident in a phrase

61. Michael Faraday, "Experimental Researches on Electricity, Third Series. [Abstract]," *Abstracts of the Papers Printed in the Philosophical Transactions of the Royal Society of London* 3 (1830–1837): 161–163.
62. Michael Faraday, "Experimental Researches in Electricity. Third Series," *Philosophical Transactions of the Royal Society of London* 123 (1833): 23–54.

– "the electric telegraph of touch"[63] – written in a tale by Longfellow in 1849. The analogy was from time to time deployed to link the body with a social system laced with communication media, which reflected the growing dependence of administrations on the telegraph. Clevenger, for example, wrote, "Sociologically ... Telegraphs and other such means of communication constitute the nervous system."[64] And in a popular lecture the electrophysiologist du Bois-Reymond explained

> just as the central station of the electrical telegraph ... is in communication in Berlin with the outermost borders of the monarchy through its gigantic web of copper wire, so the soul in its office, the brain, endlessly receives dispatches from the outermost limits of its empire through its telegraph wires, the nerves, and sends out its orders in all directions to its civil servants, the muscles.[65]

While du Bois-Reymond exploited the analogy to explain features of the nervous system, a complementary analogy described the British telegraph system. The telegraph system was centred behind the façade of a telegraph office in London "between lofty houses" which was described as a "narrow forehead" that concealed

> the great brain ... of the nervous system of Britain, [and] beneath ... the alley lies its spinal chord, [which] transmit[s] intelligence as unperceived as does the medulla oblongata.[66]

Some analogies went into a great deal of technical detail. Spencer in his elaborate comparisons of social systems and the body saw the

63. Henry Wadsworth Longfellow, *Kavanagh A Tale ....* (Boston: Ticknor, Reed, and Fields, 1849), 129.
64. S. V. Clevenger, "Natural Analogies," *The American Naturalist* 26 (1892): 195–210.
65. Emil du Bois-Reymond, *Ueber thierische Bewegung, reden* (Berlin: von G. Reimer, 1851) 29. Translated by Laura Otis in "The Metaphoric Circuit: Organic and Technological Communication in the Nineteenth Century," *Journal of the History of Ideas* 63 (2002):101–128.
66. Andrew Wynter, "The electric telegraph," *Quarterly Review* 95(1854):118–164.

telegraphic channels for communication analogous to the nervous system including the sheathing that insulates one telegraph wire from another.[67]

George Prescott, in his book about the telegraph, brought together elements of social organisation, an electrical telegraph and the nerves in a quite specific context. He associated a municipal fire-alarm telegraph system with "the nervous organization of the individual". He explained that the analogy did not guide the building of the fire alarm system but once built the analogy with "the laws of individual life" illustrated "the correspondence of the system to natural law."[68] That is, the analogy through its link with Nature's representative, the human body, provided a justification for what had been built.

Technological change led to a change in what was exploited as an analogy so it became "the destiny of the telephone" rather than the telegraph to provide "a national nervous system."[69] The increasingly elaborate technology also provided analogies that justified new neurological theories. For example, Weymouth explains

> when storms damage telephone and telegraph lines, communication can be effectively established by routes never normally used, so in the nervous system possible and efficient arcs and pathways may exist which are never normally traversed.[70]

The analogy survived the introduction of radio when it was declared the "wire and radio system of world communications [would] become the nervous system of ... human society"[71] and the British Army was claimed to have "a structure that has much in

67. Spencer, *Principles of Sociology*.
68. G. B. Prescott, *History, Theory, & Practice of the Electric Telegraph* (London: Trübner, 1860), 242.
69. Frank Baldwin Jewett, "The Social Implications of Scientific Research in Electrical Communication," *The Scientific Monthly* 43 (1936): 466–476.
70. Frank W. Weymouth, "Decerebration in Birds," *Science*, 55 (1922): 538–539.
71. John J. Carty, An unpublished address, 1923 cited in O. E. Buckley, "Medals of the National Academy of Sciences, Presentation of the John J. Carty Medal and Award to Edwin Grant Conklin," *Science, New Series* 97 (1943): 434–435.

common with the human nervous system [within which] GCHQ the brain centre, has its afferent and efferent nerves which control the entire Army."[72]

Nerves

In the nineteenth century, work on electrophysiology was continued by two significant figures who also had close associations with the growing telegraph industry and whose careers illustrate the close affinity between developments in the two fields.

Carlo Matteucci (1811–1868) studied at the University of Bologna and was later appointed as professor of Physics at the University of Pisa. He was awarded a medal by the Royal Society in London for his work on animal electricity.[73] However it was evident that he was also knowledgeable about the telegraph, since in 1850 he wrote a book entitled, *Manuale di telegrafia elettrica* and was appointed around that time to the post of director general of the Tuscany telegraph; he was dubbed "a distinguished electrician" and later became inspector-general of the Italian telegraph organisation.[74] [75]

Emil du Bois-Reymond (1818–1896), who was born and worked in Berlin, developed theories about nerve and muscle action, and was numbered, alongside Helmholtz, as one of the four bright stars of Natural philosophy. Reymond himself published the first volume of his book on *Investigations into Animal Electricity* in 1848 and although his major contributions were in the field of neurophysiology he was familiar with developments in the field of telegraphy through a number of acquaintances including Helmholtz and (Ernst) Werner Siemens, who invented an electrical telegraph system that was exploited commercially.[76]

Progress in exploring nerve action was, however, slow and

72. Anon, "Army Signals Test An Active Day In Surrey. Large Scale Operations," *The Times*, 43740, August 26, 1924, 8, col. B.
73. Anon, "Royal Society," *The Times, 18784*, December 3, 1844, 3, col. C.
74. T. P. Shaffner, "Communication with America, viâ the Faröes, Iceland, and Greenland," *Proceedings of the Royal Geographical Society of London* 4 (1859–1860): 101–108.
75. William Fox,*The Catholic Encyclopedia*, Volume X (New York: Robert Appleton Company, 1911).
76. Gabriel Finkelstein, "M.duBois-Reymond goes to Paris," *British Society for the History of Science* 36 (2003): 261–300.

development of electrical technology shifted towards general and commercial applications. In 1895 Pupin, in an address to the New York Academy of Sciences, reviewed electrical research and failed to refer to any work on nerves and muscle.[77] Even by 1912 Keith Lucas was not able to be specific about the action of the nerve. All he could say was "A change of unknown nature travels along the nerve". In a dispirited opening to his Croonian lecture, he admitted that half a century earlier physiologists must have thought they were close to revealing "the physico-chemical nature of the nervous impulse." Yet, he remarked, "the problem is still unsolved."[78]

Ultimately, the electronic valve – originally called the audion and applied to radio communication systems – and its ability to amplify rapidly changing electrical signals enabled Edgar Adrian to make significant advances.[79] [80] He explained

> The signals which [nerves] transmit can only be detected as changes of electrical potential, and these changes are very small and of very brief duration. It is little wonder therefore that progress in this branch of physiology has always been governed by the progress of physical technique and that the advent of the triode valve amplifier has opened up new lines in this, as in so many other fields of research.[81]

Adrian concluded "the history of electrophysiology [had been] decided by the history of electric recording instruments"[82] but

77. M. I. Pupin, "Tendencies of Modern Electrical Research," *Science, New Series* 2 (1895): 861–880.
78. Keith Lucas, "Croonian Lecture: The Process of Excitation in Nerve and Muscle," *Proceedings of the Royal Society of London. Series B* 85 (1912): 495–524.
79. Lee de Forest, "The Audion – Detector and Amplifier," *Proceedings of the Institute of Radio Engineers* 12 (1914):15–36.
80. Bennett, "The early history of the synapse: From Plato to Sherrington."
81. Edgar Adrian, Nobel Lecture, December 12, 1932, in *Nobel Lectures, Physiology or Medicine 1922–1941* (Amsterdam: Elsevier Publishing Company, 1965).
82. Edgar D. Adrian, *The Mechanism of Nervous Action: Electrical Studies of the Neurone* (Philadelphia: University Pennsylvania Press, 1932), 2. Cited in Robert G. Frank Jr., "Instruments, Nerve Action, and the All-or-None Principle," *Osiris, 2nd series* 9 (1994): 208–235.

such instruments had themselves relied on developments in other branches of electrical science such as telegraphy.[83] [84] Electrophysiologists did introduce refinements to instruments and their dependence on electrical engineering did establish a familiarity that spilled over into the coining of analogies. But it became evident that the electrical conduction exploited in communication systems was different from that in neurons and that the relationship between neurons and wired communication would remain that of an analogy.

Brains

Electronic instruments became more sophisticated, more was found out about the neuron and its distinctive mode of conduction and attention shifted towards examination of chemical changes.[85] It was also becoming apparent that individual neurons could provide more than a simple connection. The old analogies seemed superficial or irrelevant so analogies tilted towards ever greater complexes of both electronic devices and neurons.[86] It was proposed that "brains are less like levers and gears than like radar and thermostats"[87] and eventually the development of the electronic computer encouraged the exploration of "the analogy between the brain and the complex computing machines of modern electronic engineering."[88]

Earlier the telegraph engineers would talk about electricity flowing in the wires while the telegraph's customers saw messages that had a personal resonance. The engineers were concerned with the creation and propagation of signals and the users in what can be signified. Neurophysiologists, delving into how the neurons worked and taking the engineers' stance, reported the discovery that nerves and muscle fibres have an all-or-none characteristic

83. Ernst-August Seyfarth, "Julius Bernstein (1839–1917): pioneer neurobiologist and biophysicist," *Biological Cybernetics* 94 (2006): 2–8.
84. L. A. Geddes, "Contributions of the Vacuum Tube to Early Electrophysiological Research," *IEEE Engineering in Medicine and Biology* 20 (2001): 118–126.
85. Erwin Neher, "Ion Channels for Communication Between and within Cells," *Science* 256 (1992): 498–502.
86. Walter M. Elsasser, "A Reformulation of Bergson's Theory of Memory," *Philosophy of Science* 20 (1953): 7–21.
87. J. O. Wisdom, "The hypothesis of cybernetics," *British Journal for the Philosophy of Science* 2 (1951): 1–24.
88. Elsasser, "A Reformulation of Bergson's Theory of Memory."

and this seemed to chime with 0 and 1 used in the description of the developing electronic digital circuits and the true and false of mathematical logic. The complex electrochemistry of the nerve, for some purposes, could be ignored and, as McCulloch and Pitts claimed, "the 'all or-none' character of nervous activity is sufficient to insure that the activity of any neuron may be represented as a proposition"[89] which would enable it to be described by "the Boolean algebra of propositions."[90] These associations led to the conclusion that "electrical circuits which are used in electronic computing machinery seem to have the essential properties of nerves"[91] and an analogy was established between neurons and electronic gates – the analogues of operations in mathematical propositions. In summary, it was observed "that in electronic machines an impulse is either received or not received, a 'gate' to a circuit is either open or closed, i.e. they work according to the 'all or none' principle; and the same is true of the nervous system."[92]

Sometimes the analogy was elaborated, so for one author both the brain and the electronic computer were "considered as extremely complicated combinations of transmission lines, gates, and rectifiers",[93] but these intricate analogies were rare. In treating electronic circuits as analogues of mathematical propositions all kinds of detail could be swept aside. For instance, in the analogy, the energy supply to operate the logic gates is ignored. There are, therefore, features of an electronic circuit that have no parallel in analogous Boolean logic statements so, by focusing on the vocabulary and grammar of Boolean logic, the analogue hides some complications from view. In biochemistry, the electronic gate or the Boolean proposition furnished analogies and hence abstrac-

89. Warren. S. McCulloch and Walter Pitts, "A logical calculus of the ideas immanent in nervous activity," *Bulletin of Mathematical Biophysics* 5 (1943): 115–133.
90. Warren. S. McCulloch and Walter Pitts, "The Statistical Organization of Nervous Activity," *Biometrics* 4 (1948): 91–99.
91. A. M. Turing, "Intelligent Machinery," in *Cybernetics: Key Papers*, eds. C. R. Evans and A. D. J. Robertson (Manchester: University Park Press, 1968), 26–53.
92. Wisdom, "The hypothesis of cybernetics."
93. E. Manier, "Functionalism and the Negative Feedback Model in Biology" in *Proceedings of the Biennial Meeting of the Philosophy of Science Association* (1971) 225–240.

tions that supplied literary devices for structuring and simplifying biochemical text so, for instance, for one pair of authors, "a Boolean gate representation can characterize biochemical repression or activation of transcriptional promoter elements."[94]

Systems

For some, even the 0 and 1 of Boolean logic is a mere detail and the field of cybernetics took abstraction to a new plane, abstracting away detail until different kinds of entities could be talked about in the same terms. In the body, this strand of development drew not on individual neurons, but on neural "circuits" that, for instance, controlled "the velocity of movement, to keep it smooth and in constant relation to moving physical objects [while] [o]thers keep the body at even temperature, the blood pressure constant, and the respiration sufficient to hold the carbon dioxide and oxygen tensions at proper values."[95] Some systems being created by engineers were seen as being analogous and both the biological and engineered systems were assumed to rely on a common structural feature, feedback. This recognition was expected to allow the "design of control mechanisms which are electromechanical analogues of the human neuromuscular system."[96]

Although industrial plant seems to be on a different scale to the body, the analogy was still employed, and oil refineries, copper mines and steel mills were regarded as having "new motor and sensory nerves" making them susceptible to centralised control.[97] And this suggested that machines could replace people in an evolutionary process that supplanted muscle, simulated the nervous system and imitated mental powers.[98]

Abstract notions of communication became the common ground so advanced communication systems could be compared with nerves but now as a means for sending whole messages. For

94. Harley H. McAdams and Lucy Shapiro, "Circuit Simulation of Genetic Networks," *Science, New Series* 269 (1995): 650–656.
95. McCulloch and Pitts, "The Statistical Organization of Nervous Activity."
96. J. F. Reintjes, "The Intellectual Foundations of Automation," *Annals of the American Academy of Political and Social Science* 340 (1962): 1–9.
97. William H. Norton, "The Social Service of Science," *Science* 13 (1901): 644–654.
98. Reintjes, "The Intellectual Foundations of Automation."

instance, one author claimed, the "transmission of messages by means of a sequence of pulses is a comparative novelty in communication engineering, but in biological evolution it is as old as the nervous system".[99] Fifty years before, in the opening of his Nobel speech given in 1932, Edgar Adrian asserted:

> The sense organs respond to certain changes in their environment by sending messages or signals to the central nervous system. The signals travel rapidly over the ... sensory nerve fibres, and fresh signals are sent out by the motor fibres to ... the appropriate muscles.[100]

In doing so he provides a description close to that offered by Phineas Fletcher another three hundred years earlier. In 1633 Phineas Fletcher, who, using geography as an allegorical framework in a poem about the mind and body, represented the activity in the nerves by a technology from his era, a river carrying vessels (poasts) for transporting messages. He wrote, for instance,

> For when the Prince hath now his mandate sent,
> The nimble poasts quick down the river runne,
> And end their journey, though but now begunne;
> But now the mandate came, & now the mandate's done.

Then in a marginal note Fletcher commented, "[T]he nerves [convey] sense and motion from the brain. The will commands, the nerve brings, and the part executes the mandate, all almost in an instant."[101] Thus the Prince and the brain were seen as centralised controllers of action.

While these analogies are plausible, Adrian's "messages" and Fletcher's "poasts" affirm nothing about the detail of what is conveyed. The shift to an analogy creates an abstraction that leaves a vagueness to be filled in by the connotations conjured up by the

99. Manier, "Functionalism and the negative feedback model in biology."
100. Adrian, "Nobel Lecture."
101. Phineas Fletcher, *The purple island, or, The isle of man together with Piscatorie eclogs and other poeticall miscellanies* (Cambridge: Universitie[sic] of Cambridge, 1633), Cant.2, 20, para.13, Note k.

reader. The analogy also transforms the setting in which moral and ethical judgements are to be made and potentially displaces significant moral issues.

Around 1948 the term cybernetics was coined by Norbert Wiener who was trying "to find the common elements in the functioning of automatic machines and of the human nervous system". He aimed "to develop a theory which will cover the entire field of control and communication in machines and in living organisms."[102] At the heart of work on cybernetics were ideas of signals, communication and feedback,[103] and it was claimed that, "Wiener's efforts were devoted to effacing the distinction between human and machine."[104] So the aim was to find a neutral territory, an abstraction that could seamlessly provide an analogy for describing both machines and people. It offered a context-independent vocabulary that was solely about connections, structures and signals. As with all analogies it suppressed detail yet induced, invited and directed the imagination and delegated responsibilities for elaborating any detail to the reader.

Biological Epilogue

With a language of systems and signals in place, processes in the body and outside, for instance, the endocrine system, began to be treated as communications systems.[105] Bacteria, it was contended, "produce extracellular signalling molecules that are responsible for the communication between bacterial cells"[106] and hormones, growth factor proteins and neurotransmitters were described as signalling agents.[107]

The chemistry of living tissue had been described in mechanical terms by treating proteins as though they had "moving parts

102. Norbert Wiener, "Cybernetics," *Scientific American* 179 (1948): 14–19.
103. Wisdom, "The hypothesis of cybernetics."
104. Peter Galison, "The Ontology of the Enemy: Norbert Wiener and the Cybernetic Vision," *Critical Inquiry* 21 (1994): 228–266.
105. F. H. George, "Machines and the Brain," *Science* 127 (1958): 1269–1274.
106. Mark H. J. Sturme et al., "Cell to cell communication by autoinducing peptides in gram-positive bacteria," *Antonie van Leeuwenhoek* 81 (2002): 233–243.
107. Bernard Perbal, "Communication is the key," *Cell Communication and Signaling* 1 no.3 (Oct 27, 2003), http://www.biosignaling.com/content/1/1/3.

like hinges, springs, and latches" and structural elements "like bricks or ropes or wires,"[108] but the treatment of biological systems as communication and information systems changed the grammar and it became possible to write, for instance, that heredity is "about the transmission of information, and not just matter"[109] and that "inheritance is a discourse through time."[110] In the processes of life, it was proposed, "DNA is a molecule that stores many programs that 'compute' proteins …and the computing takes place inside molecular machines."[111] A genome became "a Read–Write information storage system"[112] and cells were treated as "complex information processing units that respond … to environmental and internal signals."[113] Parallels were drawn between the components of such systems, the "genetic switching circuits", and "electrical switching circuits"[114] and "instead of electrical signals representing streams of binary ones and zeros, the chemical concentrations of specific DNA-binding proteins and inducer molecules act as the input and output signals of the genetic logic gates."[115] Thus "biomolecular computing" was performed by "elementary building blocks, analogous to electronic logic gates"[116] and, it was presumed, "biological circuits use chemical diffusion to carry data from one component to the next."[117] Feedback too had a role so it

108. Hans Moravec, *Mind Children – The Future of Robot and Human Intelligence* (London: Harvard University Press, 1988), 72.
109. John Maynard Smith, "The Idea Of Information In Biology," *The Quarterly Review of Biology* 74 (1999): 395–400.
110. S. Jones, *The Language of the Genes* (London: Flamingo, 2000). Cited in Francesco Bernardini, Marian Gheorghe and Mike Holcombe, "PX systems = P systems + X machines," *Natural Computing* 2 (2003): 201–213.
111. Bennett Daviss, "The Computer Within," *New Scientist* 2220 (2000): 18–22.
112. James A. Shapiro, "Thinking about evolution in terms of cellular computing," *Natural Computing* 4 (2005): 297–324.
113. Subhayu Basu, David Karig and Ron Weiss, "Engineering signal processing in cells: Towards molecular concentration band detection," *Natural Computing* 2 (2003): 463–478.
114. Harley H. McAdams and Lucy Shapiro, "Circuit Simulation of Genetic Networks," *Science, New Series* 269 (1995): 650–656.
115. Ron Weiss et al., "Genetic circuit building blocks for cellular computation, communications, and signal processing," *Natural Computing* 2 (2003): 47–84.
116. Danny van Noort and Laura F. Landweber, "Towards a re-programmable DNA computer," *Natural Computing* 4 (2005): 163–175.
117. Yohei Yokobayashi et al., "Evolutionary Design of Genetic Circuits and Cell-cell Communications," *Advances in Complex Systems* 6 (2003): 1–9.

was thought that "the history of the development of feedback configurations for electronic systems may be a resource for developing an understanding of gene circuits and ... their evolution."[118]

In spite of the constant cross-referencing between electronic and biological systems, the connection with sociology had not been lost. Tangles of implications arise in analogies that coupled human societies with communities of cells; for instance when it was asserted that

> All forms of communication between human beings have long been recognized as a requirement for ... productive development of societies. This also applies to living cells who are organized in 'microsocieties' that constantly adjust to their environment through a complex network of signaling pathways.[119]

Thus communication infrastructures, a current technological theme, are placed at the heart of social organisation and cellular complexes, and, subtly, decentralised organisation is equated with the natural formations of cells as well as "productive development". Yet, as a result of their abstraction, the analogies shed little light on what might be communicated within the networks.

Conclusion

The long history of analogical links between the sciences of life, machinery, electrical technologies, politics and, occasionally, religion has bound them tightly, so discoveries, inventions, changes in vocabulary, theory, doctrine or technology in one domain inevitably influence discourse in the others. The answers to the question "What is life?" are consequently a reflection of contemporaneous political doctrines and pre-eminent technologies that structure accounts derived from observations about things we take to be living. Similarly our technology and our politics carry, at least in our conversations and writing, a reflected image of our discourse

118. Michael L. Simpson, Chris D. Cox and Gary S. Sayler, "Frequency domain analysis of noise in autoregulated gene circuits," *Proceedings of the National Academy of Sciences* 100 (2003): 4551–4556.
119. Perbal, "Communication is the key."

about vivification and its connection with nature. While we see in an analogy detailed connections, for example, between neurons and electronic gates or the workforce and muscle, the impact of analogy can penetrate deeper. Analogies involving, for instance, the body in communication systems carry connotations of nature into engineering, those founded on the nervous system and the brain can imply centralisation in administration, or those involving electronic devices can bring the ethics of engineering and commerce to biochemistry. Thus the dramatic power of an analogy is in its ability to infect a discourse with an alien ideology.

Currently, computer and communication networks, which became the signalling systems par excellence, provide a foil for analogy so the skin, for instance, might be treated as "an uncanny piece of engineering" that "processes immense amounts of data [and] ...sends signals to regulate blood flow, activate sweat glands, alert immune cells to marauding invaders, and block ultraviolet light". Expressed in these, terms our life-sustaining, erotically charged skin seems almost indistinguishable from the complementary analogy of the "electronic skin" donned by the earth which uses "the Internet as a scaffold to support and transmit its sensations" so that "millions of embedded electronic measuring devices... will probe and monitor cities and endangered species, the atmosphere, our ships, highways and fleets of trucks, our conversations, our bodies – even our dreams."[120] Thus through analogy, living beings metamorphose into signalling systems – signals in and signals out – and organisations and markets constitute a second nature saturated with information processors where people become networks of analogues provided by sensors in shops, hospitals, schools, streets and government offices.

Current predominant technologies in the field of computers and communication, and the analogies they inspire, are universal since they apply whatever people that send and receive messages do or say. They are simple structural analogies that often reduce accounts of labyrinthine physical, biological or social pathways to descriptions of elemental connections. The analogy reveals little

120. Neil Gross, "The earth will don an electronic skin," *Business Week* 3644, August 23, 1999: 132–134.

about the language used in any message nor the phenomenon used to convey it and conveniently sidesteps the description of the actual machinery. Such analogies have little explanatory power, conceal a great deal and omit any reference to the significance of what is conveyed. By stepping into a world of structures and machines, they create a nature that conceals moral and ethical challenges, and enter into a fantasy where the expression of moral and practical action is unduly constrained.

Often there are productive attempts to dissolve analogies as is evident in the development of theories of electricity. But in the assertion that kinship between telegraph and the nervous system "is more than similarity"[121] and in the declaration that the "nodes of the internet have begun to function as neurons"[122] there is a danger that we become so steeped in our analogies and seduced by their simplicity that analogy unconsciously becomes literal. The consequences can be that we neglect the present and the local pains and anxieties, morality and mortality, that our analogies can conceal.

121. du Bois-Reymond, *Ueber thierische Bewegung, reden* 10.
122. Gross, "The earth will don an electronic skin."

Participant Observation of an Indiscrete Coupling: Bio(techno)logical Life and Art Life at Symbiotica

Boo Chapple

Introduction

This text is an exploration of, and expansion upon, some of the theoretical and practical territory that I covered as part of a residency undertaken over the course of 2006 at Symbiotica, an art and science collaborative research laboratory in the School of Anatomy and Human Biology at the University of Western Australia. Bringing together theory and practice is never an easy task – the process by which work emerges through an iterative engagement with material and situation does not fit neatly within a *post hoc* analytic framework – but I have attempted to use this tension productively. Indeed, my central concern is what does the intersection of (the practice of) an art life and bio(techno)logical life, understood as two contemporary forms of labour, produce? How can an artist labour critically and effectively in this territory? In answering these questions, I reflect upon my experience of being in residence, the work that I produced as an outcome of the residency, and the work of several other artists whose practice engages with the life sciences. I frame this reflection within a critical discourse on the management of art, life and subjectivity within a globalised culture.

Symbiotica

Symbiotica is a unique and unusual place. A small loft, purpose built onto the top of the Anatomy and Human Biology building, it is filled to bursting with old scientific equipment, boxes of tissue culture paraphernalia, dismembered art installations, and other miscellaneous items. At the time of my residency it was the only art/science research space in the world attached to a life science institution, having come into being through initial residencies undertaken by the Tissue Culture and Art Project (Oron Catts and Ionat Zurr) within the School. Using this space as a base, artists are able to work in the specialist laboratories within the School of Anatomy and Human Biology in addition to accessing some areas

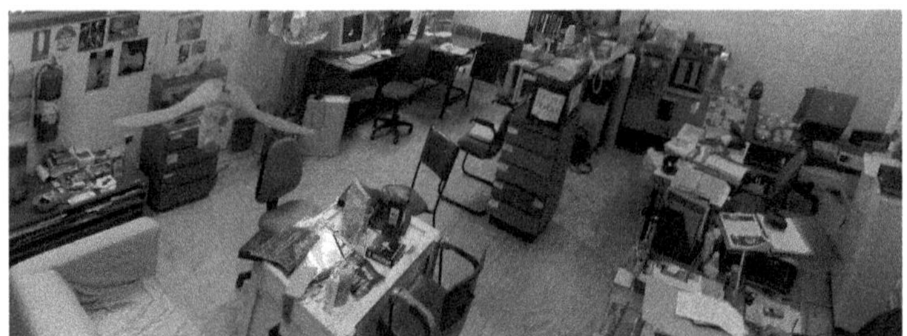

FIG. 1. Symbiotica. A panorama of the residency space at the School of Anatomy and Human Biology, University of Western Australia.

within the Faculty of Natural and Agricultural Sciences. On application for a residency, each project is assessed for its specific needs and is facilitated both by the laboratory technicians responsible for each lab and through connections that are set-up by the directors of Symbiotica[1] with relevant scientists.

Equally important to the character of Symbiotica as these physical and organisational structures, are the broader cultural parameters within which it operates, and which produce the conditions of its ongoing viability. These can be understood to arise from a convergence between two strains of thought regarding the role of the artist in contemporary society – that of the artist as interdisciplinary synthesist, whose presence is always something of an excess, a critical force, within the institution, and the 'creative industries' approach to the arts, in which the artist is invested in as an experiment in the generation of (technical) novelty and cultural capital, in this case specifically in the area of science promotion. The conflict, at once fruitful and challenging, that exists between these two perspectives is played out in the wider University framework and in the discourse around the function of arts practice within the collective identity of Symbiotica itself. The question as to why an artist should incorporate the tools and techniques of the life sciences into their practice and what is 'artful' about this work is an ongoing presence in group discussions and, indeed, it was this same question that I asked of myself. It was exciting to have access

1. When I was in residence, the directors of SymbioticA were Artistic Director, Oron Catts and Scientific Directors, Stuart Bunt and Miranda Grounds. However, there is now only one Director – Oron Catts.

to usually restricted technologies, equipment and expertise, but at the same time orienting myself artistically within this environment was an overwhelming task. I struggled with the tension between being a critical voice – using the tools at hand to express something of the contemporary condition in which biological life is at stake – and between making things that (technically) *worked*. In hindsight, I can see that these feelings of disorientation were a functional part of the experiment – Symbiotica is a space in which both biological life and art life are problematised, where the performance and construction of the artist's subjectivity become as much a part of the work as any intervention into the fabric of the life sciences.

Biotechnical Reproduction

Artists who come to Symbiotica come with many different types of projects, but the most strongly represented thematic is an engagement with political and ethical questions surrounding contemporary biotechnologies. In order to anchor myself within this discourse and forge a necessary relationship between what I was doing in the lab and the need for an art outcome, I spent time thinking through how some theoretical perspectives on art and biotechnology might be seen to intersect. One important point of departure in this respect was Walter Benjamin's classic essay *The Work of Art in the Age of Mechanical Reproduction*, my question being: How do these forms of mechanical reproduction to which Benjamin refers relate to modes of (re)production in which biological life is implicated? For Benjamin, the mechanical reproduction of images, particularly in the form of photography and film, operates to diminish the authenticity, or 'aura', of the original work of art, or subject of representation, to "detach the reproduced object from the domain of tradition" and substitute "a plurality of copies for a unique existence."[2] This lends itself increasingly to an art that is "designed for reproducability"[3], in which there is no longer any unique or authentic original. "Thus is manifested in the field of perception what in the theoretical sphere is noticeable in the

2. Walter Benjamin, "The Work of Art in the Age of Mechanical Reproduction," in *Illuminations*, ed. Hannah Arendt, trans. Harry Zohn (London: Jonathon Cape, 1970), 221.
3. Ibid., 224.

increasing importance of statistics. The adjustment of the masses to reality and reality to the masses..."[4] The work of art is now directly implicated within the practice of politics.[5] It is here that we can begin to draw a relationship to questions of biotechnical reproduction, for, as Michel Foucault begins to outline in the *History of Sexuality*, and develops in much of his later work, the modern era is one in which biological life – the very fact of being alive – becomes central to the function of the State and the management of the life, health and reproduction of a population, through the adoption of statistical techniques, becomes the purview of politics. Politics becomes biopolitics. Thus, the point at which reproduction is exposed as a technique for the mass production of reality (of nation and culture) is the point at which both art and life are brought within the domain of politics. The politics of reproduction manages both the perceptual and living labour of the masses.

Melinda Cooper, writing in her book *Life as Surplus,* outlines how the project of US neoliberalism[6] that spearheaded the 'biotech revolution' of the 1980s has been to enterprise-up this "life of the nation", to deconstruct the protective apparatus of state mediation, and expose "the realm of reproduction to the harsh light of direct economic calculus."[7] As such, if I were to define biotechnology simply, I would say that its key concern is to harness life's reproductive capacities – its capacity to regenerate, to reproduce itself, or elements of itself, in space and time – in order to realise commercial products and outcomes for the market. While it is true that living organisms have always been used to generate products for consumption, biotechnology represents something of a gestalt shift, much like the transformation enacted in the realm of art, first with the advent of photography, and now with the digital image. Indeed, in his book *Biomedia*, Eugene Thacker argues that biotechnology represents a fundamental alteration in the body-technical relation, whereby the body is no longer mediated or effected by external technologies, but rather it is itself a medium – it "generates its own technicity from

4. Ibid., 223.
5. Ibid., 224.
6. And by extension the rest of the world through US dominance in the global economy.
7. Melinda Cooper, *Life as Surplus: Biotechnology and Capitalism in the Neoliberal Era* (Seattle: University of Washington Press, 2008), 9.

within; its quality of being a medium comes first and foremost from its internal organisation and functioning."[8] [9] Thus, the fabric of 'life itself' is rationalised into a technology for the generation of capital and coupled, at an ever finer scale, to the ever expanding scope of a globalised economy, whose extent, as we know, is promulgated by a spectacular proliferation of (digital) images. Reality is reproduced as a marketable excess.

The Work of Art

So, I asked myself, what is the work of art in this context? What different excesses do artists produce? To a certain extent this is the task of the rest of the chapter to determine, but in a brief precis I would like to propose that the work of art is to activate a particular conjunction of the (economic, material, political) frameworks of biotechnology with the disciplinary structures of academic knowledge (as manifest in the institution of the University based residency) and the subjectivity of the artist – the narrative of work, life and public self presentation that is the contemporary art persona. As Benjamin so presciently described, the destruction of the aura of the work of art through mass reproduction, led in turn to the commodification of the personality behind the work as a means to re-imbue it with authority, or exclusivity.[10] Thus, the work of art in this context represents a critical nexus of the reproductive potential of bio(techno)logical life with the cultural capital of an interdisciplinary art life. While any engagement with this territory cannot help but function to incorporate both life and art within a politics of management and expose them further to the market, it is also a means by which practices of management and commodification can be questioned and redirected. For Benjamin the massification of art in the form of photography and film presented an analogous paradox. The technical-economic apparatus that operated to destroy

8. In a similar manner, for Benjamin, the technology of photography ceased to be merely a mediation and became itself an artistic medium, of which reproducability was a defining feature.
9. Eugene Thacker, *Biomedia* (Minnesota: University of Minnesota Press, 2004), 10.
10. We need only to think of the extent of contemporary celebrity culture to realise Benjamin's foresight. Benjamin, "The Work of Art in the Age of Mechanical Reproduction," 231.

the unique aura of the work of art, simultaneously functioned as a means of providing more egalitarian access to a territory historically circumscribed by class and birth. In this respect, it represented a site of struggle and the locus at which new forms of art and life were being produced. In contemporary terms, the (re)production of life and subjectivity (identity) within the advanced capitalist framework of globalisation represent just such a domain of contest and novelty. The work of art thus situates itself within this domain as a critical reproduction. It acts to draw attention to its own particular complicity in this present moment and to imagine how it may be constituted differently.

Making Strange

In thinking about how I might set about drawing attention to the complicities of my work within a broader cultural economic framework – to engage in critical reproduction – I was influenced by the writing of Russian Formalist literary theorist Viktor Schlovsky, who in his 1917 essay *Art as Technique* uses the term 'defamiliarisation', or 'making strange' to describe the way in which a work of art operates to fracture habitual relations to the world. He says:

> After we see an object several times, we begin to recognize it. The object is in front of us and we know about it, but we do not see it – hence we cannot say anything significant about it. Art removes objects from the automatism of perception… [11]

Interestingly, this essay was written in the same year that Marcel Duchamp's cannonical readymade, *Fountain,* was submitted for exhibition in the United States, at a time in which the homogeneity of industrial production – the mass reproduction of objects – was beginning to function as key determinant of society and culture around the world. As such, both articulate the need for a reframing, or redirection, of habituated mass perception – for

11. Viktor Schlovsky, "Art as Technique," in *Russian Formalist Criticism: Four essays,* translated and edited by Lee T. Lemon and Marion J. Reis (Lincoln: University of Nebraska Press, 1965), 13.

an art that makes objects 'unfamiliar'.[12] Given the foundational status of Duchamp's appropriation and recontextualisation of the 'Readymade' in the history of contemporary art, it is unsurprising that there are numerous examples of artists who take a similar approach to the objects of biotechnology. One such example, is the work *Hela* (2000), by Christine Borland, in which a pertri dish of HeLa cells[13] is placed under a video microscope in a gallery, with the image from the microscope displayed at macro scale on the wall next to the installation. However, these objects are not the everyday objects of the original 'Readymades', they are the symbols of science. They are familiar representations of the life science laboratory, re-presented in the gallery in concrete form that, though it may be three dimensional, does not subvert our habits of perception, or generate an affective jolt of the strange and singular. The particular strangeness of the conjunction between life and technology that they manifest is occluded by a symbolism that is not unfamiliar but somewhat ordinary.

At this post-industrial, post-Duchampian, intersection of biotechnology and art, 'defamiliarisation' can no longer be achieved by merely reframing, or representing, the biotechnological object in a gallery. The contemporary proliferation of the image, dissolves life into a constant re-presentation that is at once entirely familiar and, as Guy Debord so clearly reminds us, alienates us from the materiality of lived existence.[14] This makes it impossible to interrupt the habits of perception by means of a simple disjunction between object and context. Rather, I saw my own work at Symbiotica as being to create the strangeness of an unexpected *contiguity*. Defamiliarisation for me became about breaking through the habitual separation of representation to create a moment in which

12. Ibid.
13. The immortal HeLa cell line was extracted, in 1951, from a fatal tumor in an African American woman named Henrietta Lacks without her knowledge and named from the first two letters of her first and last name. The total mass of all HeLa cells now in the world is thought to greatly exceed the mass of Henrietta Lacks' body when she was alive.
14. "Though separated from what they produce, people nevertheless produce every detail of their world with ever-increasing power. They thus also find themselves increasingly separated from that world. The closer their life comes to being their own creation, the more they are excluded from that life." Guy Debord, *The Society of the Spectacle*, trans. Ken Knabb (London: Rebel Press, 2005), 33.

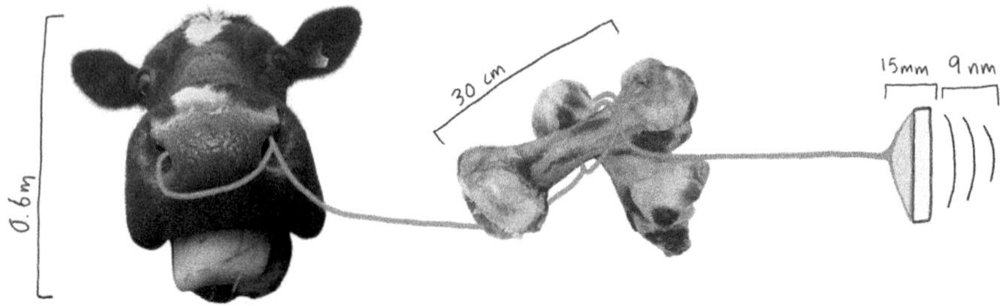

FIG. 2. Diagram depicting relationships of scale in the cow bone audio speaker process.

one is able to experience the material indiscreteness of ones own being, or the uncanny proximity between one's everyday practices and the politics of 'life itself'. My biggest challenge was how to create this sense of implication, of continuity between the everyday and the tools and techniques of biotechnology, without hitting the same wall of the laboratory aesthetic, the ordinary occult, that *Hela* and other works of that ilk represent.

Over the period of my residency I pursued this project in several different ways and with varying levels of sophistication as my own understanding of what was at stake in the process developed. First and foremost, the feature that was key to both projects exhibited as a result of the residency was a focus on materials and organisms that had a purchase outside the realm of the laboratory – bone, collagen, cows and rats. Bone represents a rich historical and metaphorical territory that stretches from ancient tools, through architecture and orthopedics and collagen is the stuff of talk shows, cosmaceuticals and face lifts. While rats are synonymous with laboratory experimentation, they are also a figure of myth and a creature of the margins. Cows populate equally children's books and the dinner table. As such, any biotechnical process that

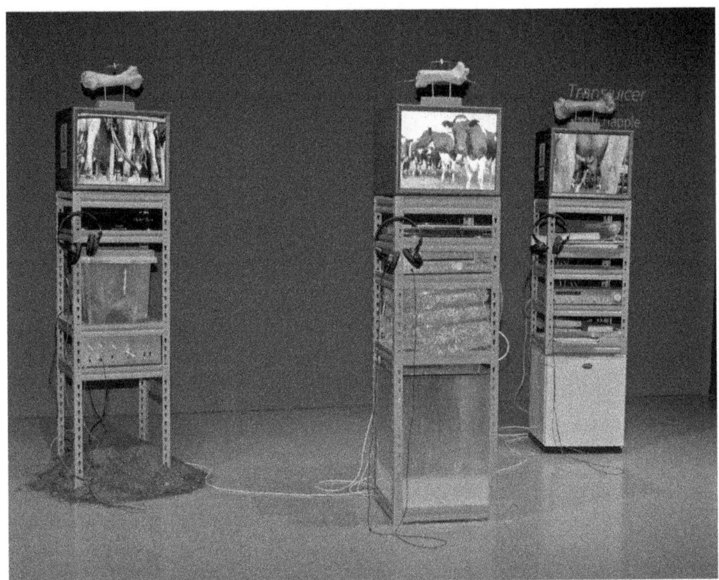

FIG. 3. *Transjuicer.*

engages these elements is necessarily continuous, materially and symbolically, with a much broader realm than that of the life sciences, and speaks to the implication of life within techniques of management – fancy rats, pest control, milking machines, the global meat market – across widely divergent scales and times. In my effort to avoid laboratory aesthetics, I attempted to draw attention to the strange capacities of familiar materials and to create connections between the transformations enacted in the laboratory with these broader significatory contexts. In this way, by virtue of association, the uses of these materials in the world outside the laboratory are also 'made strange'.

Transjuicer, a project involving audio speakers made from cow bone, uses transduction as both mechanism and metaphor to speak to this complex of symbolic and material exchange. Technically, the speakers are made by harnessing the piezoelectric nature of the bone matrix in order to cause the bone to vibrate in such a way as to generate sound[15]. At the purely material level they transduce

15. Piezoelectricity is a feature of certain crystals and polymers such that when a mechanical pressure is applied they become electrically polarised and *vice versa*. Thus if an oscillating voltage is applied to a

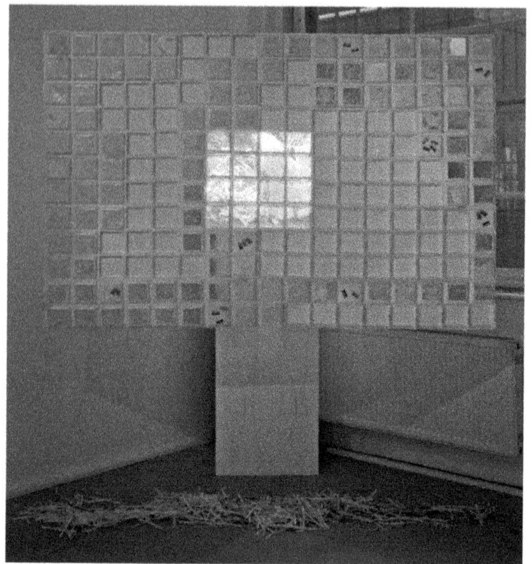

FIG. 4. *A Rat's Tale*, panel.

electromagnetic waveforms – in this case of the bone songs, cow songs, milk songs that I played through them – into sonic vibrations. These vibrations have been recorded using a laser interferometer and can be listened to on headphones. As they exhibit the strange qualities of a familiar material, they operate to fracture the habitual perception of one's own body and disrupt the reciprocal network of metaphors that inform both the way in which we experience bone and the way in which we understand things in the world that have been designated 'bone-like'[16]. At another level again, the speakers transduce between macro social context and micro technical intervention, between the cow and the gallery, between the living and the dead. Each speaker is exhibited atop a totem of transformation – a selection of artifacts from the history of the process of refinement from bloody butcher's bones to technical instrument

piezoelectric material it will vibrate at the corresponding frequency. Among other devices, piezoelectric materials are used in ultrasonic transducers, audio speakers and contact microphones.

16. We tend to use the metaphor of bone to signify something static and immutable – the deep structures of architecture and logic, for example.

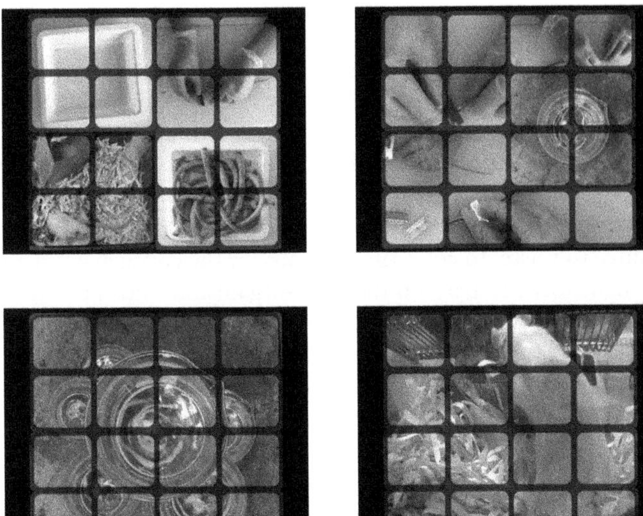

FIG. 5. *A Rat's Tale*, videostills.

– and accompanied by a video piece shot at a local dairy farm that documents the intimate interface between cow and (milking) machine already existing at the heart of our everyday consumption practices. As such, they draw together the abstraction of biological materials in the laboratory with the trajectory of instrumentalisation that begins with the domestication of the cow itself.

A Rat's Tale is comprised of a video projection onto a panel constructed of laboratory weighing trays and pieces of dry collagen extracted from 'waste' rat tails – tails that I collected from rats that had been sacrificed for research. The trays and collagen are patterned together into the layout of the first maze ever used in experiments conducted with rats. The video, composited from documentation of the process of material transformation from living rat to collagen fragment, is projected at the centre of the maze. The panel itself is suspended on lengths of steel cable that hang curling like tails towards the shredded paper spread on the floor below. While this work is more obviously framed within the laboratory context than *Transjuicer*, it is constructed in such a way as to draw attention to what is around the edges and in the cracks of the scientific process. The white panel is not the wall of the science communication laboratory aesthetic. Rather it is fragile and porous to the world and to the mess of life within which it is entangled. The video begins and ends in a continuous loop with footage of the rats themselves living in their shredded paper nests – replenished weekly with the discarded paperwork of the bureaucracy that manages their passage through the world from birth to death, from resource to waste. In collecting the tails for the collagen extraction, I brought focus to bear on the extra-organismic ends produced in the process of bio(techno)logical research other than those that finish up in the publication or at the market, in the excess that could be harnessed in the critical reproduction of art. I investigated the potential of using scientific techniques to produce something that did not look like science – a mirror to its ordinary face. I also set out to draw attention to the strange and abject practice of producing collagen, a beautification agent injected below the surface of many a human body, from such cadaverous waste products[17].

As a result of my investment in making work that was not solely focused on the action of a technology, neither of these works operate on the terrain of what one might consider to be biotechnology proper, in which the reproductive capacity of 'life itself' becomes a technology, a medium manipulated and enterprised up at the micro scale (and beyond) for the generation of commercial products and outcomes. The extra-organismic ends – artifacts, sound, video – were produced differently for the purpose of art, as much through my performance of the process, extracted and re-presented as a material-narrative excess, as through the generative capacities of rats and cows. Both works are designed in such a way as to draw attention to the iterative feedback loops between life and non-life, organism and institution, animal and image that continually reproduce the complex of economic, bureaucratic, and symbolic relations that represent, in turn, the incorporation of reproduction within networks of management and capital – from the knowledge industry of the life sciences to the machinations of global food production. The artifacts produced and extracted in the lab, form only a small part of the work and might better be under-

17. Though in this case it is usually extracted from cow skin and tendon, not from rat tails.

stood as the initial departure point from which my complicity as an artist, and by extension the complicity of the audience, in this complex of relations is explored, opened up, and 'made strange'.

Performance and Pedagogy

The performance of the artist within the institution, in engagement with the materials and tools of biotechnology, and within the art and academic context of re-presentation – the narrative of the art life persona – is pivotal. Indeed, in a contemporary echo of Benjamin, Miwon Kwon, writing about the condition of 'itinerant artists' in the global art economy, sees that " the artist comes to approximate the 'work'".

> It is now the *performative* aspect of an artist's characteristic mode of operation (even when working in collaboration) that is repeated and circulated as a new art commodity, with the artist him/herself functioning as the primary vehicle for its verification, repetition, and circulation. [18]

Symbiotica is a site at which these performances take place and its name is repeated and authenticated by a host of international artists as they spread out and circulate themselves, and the outcomes of their work, through a network of exhibitions, symposia, and publications. It is unsurprising that this combination of global performativity with the new media of the (biotech) body that Symbiotica represents has attracted older generation performance artists such as Orlan and Stelarc, as well as more contemporary performance practitioners, to undertake residencies and collaborative projects at the space. The labour of art in this contemporary context becomes one of performative reproduction. The art *is* the reproductive potential of an art life in the same way that the technology of biotech *is* the reproductive capacities of the biological. The art object (when there is one) has the power of the strange, or singular, only in so far as it is able to reproduce a set of material, biological, narrative relations that speak to this performance.

Given the necessity of an institutional framework to support

18. Miwon Kwon, *One Place After Another: Site Specific Art and Locational Identity* (Cambridge, MA: MIT Press, 2004), 47.

the work of an artist whose practice involves a hands-on engagement with the tools and techniques of biotechnology, it is not unusual that the performance of this kind of art should regularly take a quasi-academic, or pedagogical, form. The necessary presence of the artist, as principal interpreter, with the artwork, or the function of the artwork as a pedagogical tool with which the artist performs, are common tropes, as are the lecture, the workshop, or the publication (such as this one), by which the artist disseminates their theoretical and/or technical narrative and produces authority for their work. At Symbiotica, this plays out through lectures and events that take place in and around the School as well as through an extended network of connections between institutions around the world. Artists in residence are expected to give a presentation on their work as part of the School of Anatomy and Human Biology lunchtime lecture series and also attend and present at the weekly Symbiotica 'lab meetings'. The founders of Symbiotica, The Tissue Culture and Art Project, are regular guests at conferences and events both locally and internationally and have forged a name for themselves at the intersection of art, design, bioethics and the cultural studies of science. The exigencies of a highly technical practice, involving the maintenance of their 'semi-living' tissue sculptures, mean that the work cannot easily be shipped and installed without their presence. As such, the installation process, the producing of the work *in situ*, at the point where the reproductive force of art life and bio(techno) logical life meet, is very much a performance, which they use to engage both the art audience and the art institution in a pedagogical experience of the biotechnical and ethical issues at stake in the practice of tissue culture. In a similar manner, Paul Vanouse, another former Symbiotica resident, uses his work with DNA gel electrophoresis as a platform to give lecture-performances about the constructed nature of the DNA fingerprint. The act of running the gels to generate the two-tone bitmap images of *Latent Figure Protocol* is a critical reproduction, in which the processes of life and science are redirected and leveraged as a tool to educate a public regarding the politically invested nature of these same techniques.

While my own works do not incorporate a live performance, nor are they fed or cared for over the course of the event, they are, nonetheless, most often exhibited in a context that involves a thematically related symposium where I present an accompanying artistic-academic discourse. In addition to this, the artifacts and video of both *Transjuicer* and *A Rat's Tale* were constructed to operate as testaments to the original performance – to amplify certain aspects of this material-institutional intervention for the experience of the audience. For example, the process of extracting the collagen for the panel in a *A Rat's Tale* was an amplification performed through iteration, a testament to the scale of rat usage in research and to the time invested in the extraction. A simplified set of the repeated actions that I performed to extract the collagen – cut, separate, recombine – were applied, in turn, to the laboratory weighing trays to create the panel. The paper that I had used in storyboarding the video, research notes, and glove boxes were all put through the shredder just as the shredded paper of the animal house administration was fed back into the task of raising more rats, and just as the rats themselves were dissected to feed the need for research outcomes. The final piece was, thus, a reproduction through disarticulation and recombination. The performance of my art life was fed back into the process as a commentary on my own complicity in this story of a rat.

Beyond the immediate context of the art event, I find that the majority of my work and the outcomes of the residency are not to be found in the art objects at all, but in the institutional, interpersonal, interdisciplinary and pedagogical relationships and competencies that have developed as a result. The act of exhibiting and presenting the work becomes one of reproducing, consolidating, and extending upon these relations. In working at Symbiotica the task of collecting the rat tails lead me to develop a network of relationships with researchers in the School and a knowledge of the operations of the animal facility, such that when I moved back to RMIT University I was easily able to access their animal facility and an ongoing supply of 'waste' tails. The final work, is thus as much a production of tacit disciplinary knowledge and institutional competency as it is of lab rats and collagen extraction. Equally, the work that I did on the bone audio speakers was as much an artifact of a pedagogical process as it was a material transformation. In order to obtain access to the equipment and expertise required to undertake this work, I took on the co-supervision of a Bioengineering Honours student (William Wong) who worked with me throughout the year. In this respect, the performance of art becomes one of service to a set of broader institutional needs where the role of the artist is both managed and manager – an instrumental function in the reproduction of institutional forms and subjectivities.

This question of the performative reproduction of institutional forms through the process of art is one dealt with extensively by George Yúdice in his book *The Expediency of Culture*, where he demonstrates how art in the era of neo-liberal globalisation becomes increasingly distributed through the social fabric as a cost effective mechanism of 'civilisation' and the development of spaces for capital accumulation.[19] The presence of Symbiotica and its resident artists in the School of Anatomy and Human Biology functions in this way to provide an outlet for discussion and a social space for certain scientists and graduate students, as well as to bring people together from other parts of the University and from the wider arts and intellectual community. The interdisciplinary excess thereby produced is a flagship for the University, serving to advertise its commitment to the production of novel forms of knowledge and community that are central to maintaining a competitive edge in the contemporary higher education market. For the artists who are charged with producing the concrete evidence of this excess, it is as much a process of being disciplined as it is of being interdisciplinary. Conducting research into scientific methods and techniques and particularly working in the laboratory means being formed to the requirements of an academic discipline and the affordances of the devices and procedures of the lab. In the production of an outcome, the capacity of our bodies to perform and reproduce that performance consistently, were as significant to the final artifacts as the capacity of the biological materials with which we were working and the institutional framework through which we operated.

19. One instructive quote in this regard that Yúdice takes from *American Canvas,* a 1997 report from the National Endowment for the Arts: "No longer restricted solely to the sanctioned arenas of culture, the arts would be literally suffused throughout the civic structure, finding a home in a variety of community service and economic development activities – from youth programs and crime prevention to job training and race relations..." Quoted in George Yúdice, *The Expediency of Culture: Uses of Culture in the Global Era* (London: Duke University Press, 2003), 11.

Conclusion

A residency at Symbiotica is one in which both art life and bio(techno) logical life are called into question. This is a disorienting process from which I have attempted to unravel the narrative threads of an argument that addresses the multivalence of 'reproduction' as it is engaged in this particular economy of art. In thinking through how a critical reproduction can be enacted, I have brought my attention to bear on the iteration of performative action and the implication of this action within certain institutional and economic frameworks engaged in the contemporary management of life. I have found it useful to reflect upon the indiscreteness of our lives with respect to the material and institutional operations of a globalised culture. What does it means to be both observer and participant, producer and product, in this economy? How can we continue to work towards both the dehabituation of an audience and our own dehabituation to the actions in which we are engaged? Perhaps bearing witness, in this way, to these questions and the complex entanglements through which a life, my life, is performed is a start:

> to witness an event is to be present at it in some fundamentally ethical way, to feel the weight of things and one's own place in them, even if that place is simply, for a moment, as an onlooker.[20]

20. Tim Etchells, *Certain Fragments: Contemporary Performance and Forced Entertainment* (London: Routledge, 1999), 17.

Building Better Beef: Biotech and the Construction of Cattle

Ron Broglio

> Man can hardly select, or only with much difficulty, any deviation of structure excepting such as is externally visible; and indeed he rarely cares for what is internal.
> Charles Darwin, *Origin of the Species.*[1]

Cattle are neither simply part of nature nor elements of culture. The body of the animal has been changed by human choices in selective breeding designed to meet human desires for more beef and tastier cuts of meat. As such, the cattle body is part of culture. Yet, in their animality, the cattle stubbornly remain part of nature. As they graze on humanly planned fields and feed lots for the end purpose of human consumption, these animals of nature and culture are hinge objects that work between the realms of the artificial and the natural, with each jostling in relation to the other. Humans want the very body of the animal but only as already codified parts of meat. Over the last two hundred years we have improved the breeding of cattle by an increasingly careful selection of animals according to an ideal. This ideal appears in beef diagrams that show cuts of meat and is manifest in tests for tenderness and "marbling" or intramuscular fat. Such representational models and calibrations of animal parts get projected onto actual animals who are called upon (culled and bred) to become increasingly ideal beasts. The representational-then-materialized ideal animal translates the actual cattle's animality into a literally digestible form for circulation within culture's consumption. Of interest in this essay are the technological means by which representations of cattle become actualized upon physical animals and the resulting effects on the breeds and species. Technology increasingly intrudes upon any space that cattle as cattle may have outside of human culture. By bringing the body of the animal and the interior(ity) of the animal into the folds of culture, the animal's space unto itself is lost. This essay seeks to trace the complex slippage between nature and arti-

1. Charles Darwin, *Origin of the Species* (New York: Gramercy Books, 1979), 96.

fice in cattle breeding for meat consumption. Furthermore, since "good" breeding – already a sign for socialization of beasts – entails documentable lineage, I will address issues of origin and copy, clone and original, and authentication of breed.

Animal husbandry creates mobile boundary lines between the natural and the artificial. In breeding, there is a muddled border by which cattle serve as both a tool and product, i.e. cattlemen use the body of the animal as a technology to effect change on the body of its progeny. Over several generations of selective breeding, human choices modify the animal body to better match our needs. Such a body is both "natural" in that the offspring are animals produced biologically "by" animals themselves and "artificial" in that human selection of which animals will mate has altered the form and characteristics of the cattle. Artificial insemination, embryo transfer, and now gene selection further emphasize human technological intervention, but even in its most primitive forms, selective breeding makes cattle sex and offspring "unnatural." Improved breeding practices, particularly over the last two hundred years, have perfected the cultural manipulation of the animal's body. Consequently, the animal body and the very interiority of the animal, the animal-as-animal, has come under assault by agri-culture.

Gene technologies are the latest means of building better beef and colonizing the interior of the animal. In order to develop good breeding, cattlemen read the body of the animals and mate those which have the desired qualities of good form, flesh, and propensity for fattening. By reading the body, humans begin to move cattle into the human environment. We appropriate the animal by visualizing it for our uses. The beginnings of contemporary visual appropriation can be found in 18th and early 19th century cattle portraiture. These paintings are a visual supplement to refined breeding practices. As will become evident later in the essay, genetic images extend idealization found in portraiture. In breeding, the visual ideals play themselves out on actual animal bodies. There are, then, two technologies that are intertwined – visual representations and breeding which actualizes ideals upon cattle.

Modern breeding practices can be traced back to Robert Bakewell's 18th century English farm. From his day well into the twentieth century graziers have assessed their herd by reading the outward form of the animal. The butcher's carving lines are mentally projected upon the animal and choice cattle are bred to produce ideal progeny which match grazier's projections for more meat in the right places, better "marbling" of fat, and speed by which the animal can achieve these ends. In brief, animals are chosen for form, flesh, and propensity for fattening. However, in the last few decades gene technologies have changed what is possible for breeding. Traditionally, cattlemen read the outward appearance of prized cattle, and the only glance at the animal's interior happened at slaughter when the inside of the animal becomes a visible outer surface of meat. In the case of prized breed cattle, the animal itself would not be slaughtered for years; instead, its traits were discerned outwardly and by the meat produced by its progeny. However, today with functional genomics the outer surface of cattle hair and skin are transformed into genetic information which yields a look "inside" the beast. Given this look inside, computer programs which run genomic inputs now allow cattlemen to match theirs cow and bulls to achieve "targets" or ideal offspring. Currently, unlike goats and pigs, cattle are not used in transgenetics at any significant level. Rather, at the furthest ends of using genetic technologies on/in cattle, cattlemen have cloned prized breeders in order to multiply their ability to produce progeny from a single superior stock. Genetic information about cattle and cloning of cattle have introduced complex legal and social issues that highlight the continuum of technological visualization of the interior and enframing of the animal body.

Cattle have been domesticated for roughly 9,000 years and as such have been a part of culture. Within this long relationship between cattle and humans there are several decisive moments which signal the increased intertwining of the natural and the artificial in the body of these animals. Cattle of today (*Bos Taurus* and *Bos indicus*) evolved from the "wild cattle" called aurochs (*Bos primigenius*). The last documented wild cattle died in 1627 in the forest of Jaktorów under the possession of the Polish Royal Family. There have been some experiments in backbreeding modern cattle to try to bring back the auroch in what would be the artificial creation of a natural animal from domesticated beasts; however, it remains unclear whether scientists have achieved actual backbreeding or rather engineered a pseudo-aboriginal animal.[2] (Backbreeding raises the issues of what constitutes a "natural" end, i.e. at what point is the animal sufficiently "wild" or "natural" to end the breeding experiment and what characteristics constitute "naturalness.") As for contemporary cattle, they are fully domesticated and as such bear the markings of culture through "good breeding." Indeed, breeding practices form the second decisive moment in the cultural history of cattle. Through selective breeding we have changed the shape and disposition of the animal. Our ability to change the animal and its "insides" calls into question the commonly held idea that livestock are natural and that the meat of such animals is a natural product.

As Charles Darwin notes in his *Variations of Animals and Plants under Domestication*: "Man, therefore, may be said to have been trying an experiment on a gigantic scale; and it is an experiment which nature during the long lapse of time has incessantly tried."[3] Indeed Darwin's knowledge of domestic breeds helped him to formulate his theory of natural selection.[4] Robert Bakewell's experiments with cattle function as important examples throughout Darwin's *Variations*. Cattle breeding took a decisive turn in the 1760s when Robert Bakewell at Dishley Farm in Leicester England promoted in-in breeding (mating within the same family) to ensure that offspring would inherit the desired characteristics. By mating within what today we would call the same "gene pool" Bakewell was able to ensure the purity of desired traits and consequently he realized amazing gains in the size of his animals over a limited number of years. As many historians of Bakewell observe, farmers and breeders are slow to change methods that have been passed down from father to son over generations.[5] There are understandable reasons

2. Laurie Winn Carlson, *Cattle: An Informal Social History* (Chicago: Ivan R. Dee, 2001), 171-94.
3. Charles Darwin, *The Variation of Animals and Plants under Domestication* (Baltimore: Johns Hopkins University Press, 1986), 3.
4. Roger J. Wood, "Robert Bakewell Pioneer Animal Breeder and His Influence on Charles Darwin," *Folia Mendeliana* 8 (1973): 231-242.
5. The success of his practice is chronicled in many 18th and early 19th century publications including William Marshall's *Economy of the Midland Counties* (London: G. Nicol, 1796); George Culley's *Observations on Livestock* (New York: D. Longworth, 1804); and William Youatt's *Cattle: Their Breeds, Management, and Diseases* (London: Baldwin and Cradock, 1834).

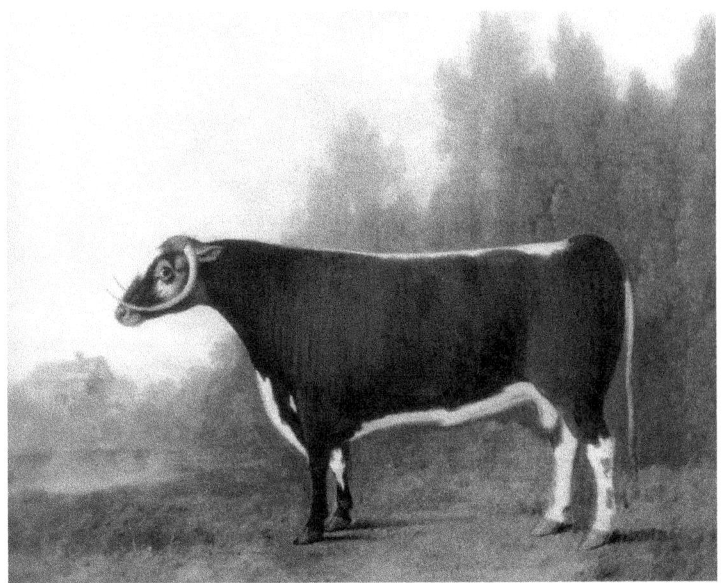

FIG. 1. *Garrick*, John Boultbee, Shugborough, Courtesy of The Anson Collection (National Trust), ©NTPL/John Hammond.

for the slow innovations in agriculture. Adopting a new practice means taking a risk with one's animals and thus one's livelihood. The number of variables are often too complex for farmers to predict; what works well in one region of the country or with one breed may not work well elsewhere or with other animals. Remarkably, within his lifetime Bakewell was able to change the British breeding practices that had remained the same for centuries. As an agricultural innovator, he focused on improving the form, flesh, and propensity for fattening of his beasts by inbreeding desired characteristics (fig. 1). Thus, animals would gain flesh and fat in less time and consequently increase the grazier's productivity and profits. Bakewell famously claimed that his animals were "the best machine for converting herbage into money."[6]

Robert Bakewell transformed the middling longhorn cattle of Leicester into a profitable animal that became the most important breed in England during the late 18th century. Unfortunately, too much in-in breeding eventually diminished the strength of

6. Harriet Ritvo, *The Animal Estate* (Cambridge, MA: Harvard University Press, 1987), 66.

his animals. In its place, the shorthorn rose to notoriety. Robert and Charles Colling made the shorthorn breed famous by drawing on Bakewell's method but also maintaining some breed virility through using "outside" stock. In 1822 Shorthorns had the first published pedigree herdbook for cattle. The breed gained strength throughout the 19th century and remains an able breed today. Creating a recognized and renowned breed is no easy matter for graziers. In the late 18th century, the rhetoric employed to define a breed shifted from qualities of provenance, function, and resemblance to a series of distinguished characteristics.[7] Since breeds are not separate species, their designation is the result of marketing a particular series of traits as distinct or "distinguished." Such marketing is valuable for gaining recognition of a cattlemen's herds. Without it, the animals are indistinct and consequently an unknown or lesser known product whose breeding characteristics and fattening qualities remain uncertain. The agricultural literature of the late 18th and early 19th century gives elaborate stories of pedigree, lineages, and description of distinguished qualities. These stories reveal how different graziers struggled to get their animals recognized as a unique herd or a distinguished breed. Throughout the pamphlets, flyers, books and journals, primacy is given to visual description of the animal and each of its valued parts. A typical description of a longhorn by the 18th century agriculturalist William Marshall reads: "His chest extraordinarily deep, – his brisket down to his knees. His chine thin, and rising above the shoulder points, leaving a hollow on each side behind them."[8] Marshall continues for another page and a half describing this prime example of British animal husbandry. Usually, such descriptions are accompanied by illustrations. Cattle images which proliferate in the 19th century served as a portable visual argument for a particular farmer's breed.

One of the fascinating elements of cattle portraiture is the way the art serves not simple as a copy of the animals but instead as a means of production. Portraits fashion an ideal toward which farmers would breed their herd. The grazier is always breeding toward something, some ideal. Chroniclers of Bakewell admire his

7. Harriet Ritvo, *The Platypus and the Mermaid* (Cambridge, MA: Harvard UP, 1997), 78.
8. Marshall, *Economy of the Midland Counties*, 323.

ability to visualize his goal of animals with more flesh in the right places. In *Variation,* Charles Darwin cites several of his contemporaries on this point:

> Few persons, except breeders, are aware of the systematic care taken in selecting animals, and of the necessity of having a clear and almost prophetic vision into futurity. Lord Spencer's skill and judgment were well known; and he writes (20/5. "Journal of Royal Agricult. Soc." volume 1 page 24.), "It is therefore very desirable, before any man commences to breed either cattle or sheep, that he should make up his mind to the shape and qualities he wishes to obtain, and steadily pursue this object." Lord Somerville, in speaking of the marvellous improvement of the New Leicester sheep, effected by Bakewell and his successors, says, "It would seem as if they had first drawn a perfect form, and then given it life."[9]

In many cases, the graziers' "perfect form" is first realized in art images. George Garrard's *Description of Oxen* serves as an example (fig. 2). It was commissioned and sponsored by the Duke of Bedford and Lord Egremont and was officially recognized by the Board of Agriculture. In his *Description*, Garrad drew an idealized version of John Day's admirable Durham Ox. Garrard went so far as to create a sculpture of this prized shorthorn for artists and farmers to study as literally a "model" animal. The role of Garrard's models, engravings, and descriptions were to provide "a sort of standard whereby to measure the improvements" of cattle both in the present and for future generations.[10] Farmers could look at the illustrations and models to understand "the ideas of the best Judges of the times" regarding points and characteristics of a breed.[11] Graziers would then buy and breed animals with the goal of matching the depicted

9. Darwin, *The Variation of Animals and Plants under Domestication*, 179.
10. George Garrard, *Description of the different varieties of oxen, common in the British Isles* (London: J. Smeeton, 1800), not paginated "Introduction."
11. Garrard, *Description of the different varieties of oxen, common in the British Isles*, not paginated "Introduction."

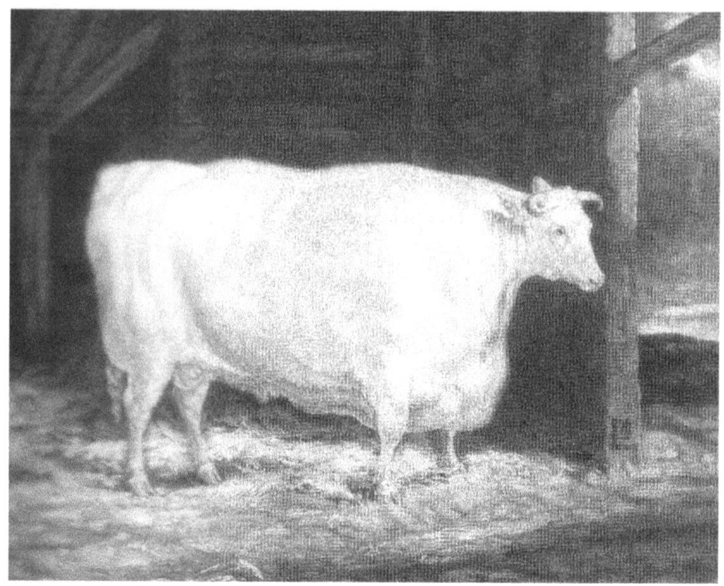

FIG. 2. *The Durham White Ox* after George Garrard engraved by William Ward, 1813. Courtesy of the Yale Center for British Art.

ideal as "first drawn . . . and then given it life." Animals begot art which engendered more and larger animals.

Since the fundamental means of evaluating and categorizing animals during the 18th century was by lines and surfaces – i.e. by what can be seen – artists have a privileged position in the promotion of breeds.[12] Distribution and hierarchy of traits and values are made through a description of what is visible in cattle. The patterns, points, and characteristics literally display themselves.[13] The same is true today in cattle shows and state fairs across the United States. One has only to look and, of course, to be aware of what one is looking for. Many of the points and characteristics of a prize bull are designed to be captured visually. It was commonly held and still often regarded that beauty of form creates the finest

12. Michel Foucault, *The Order of Things* (New York: Vintage Books, 1994), 137.
13. A knowledge about animals based on external visible surfaces fits with the Linnean classification of flora and fauna. For a more detailed look at the history of naming and classification based on animal exteriors during the Early Modern period, look at the chapter "Seeing All Their Insides: Science, Animal Experimentation and Aesop" in Erica Fudge's *Perceiving Animals* (Urban IL: University of Illinois Press, 2002): 91-114.

animals.[14] With a primacy placed on line and form by breeders, the artist is able to employ his skills toward creating a visual argument for the value of an animal. In painting, the cattle's quality becomes evident at one glance.

While embellishing ideal characteristics, animal portraiture essentially repeats the visual surface of the animal in the field. It is a repetition with a difference marked by the painter's idealization of cattle traits. As these illustrations get realized in the flesh of animals, the cycle of illustration and breeding spirals toward a teleological more fit carcass of meat. In this process, the being of actual animals is reduced to a morphological becoming as breeders position these beasts on the ever extended horizon of better beef. While quite obvious and perhaps unspoken, breeding changes the physiology of animals, their body and comportment in the world. In pursuit of better beef, the 19th century shorthorn cattle breeder John Wright notes that "In the anatomy of the shoulder modern breeders have made great improvement on the Ketton shorthorns by correcting the defect in the knuckle or shoulder-joint, and by laying the top of the shoulder more snugly in the crop, and thereby filling up the hollow behind it."[15] This modification through breeding adds more meat to the animal and supplies a steady shoulder upon which the hefty carcass can rest. There exists quite a litany of such changes in the history of each breed. Yet as breeders improve cattle and yield better meat, they are hindered by their inability to see inside the animal in order to base mating selection on the quality of the animal's flesh. Instead, cattlemen make their decisions based on reading the exterior of the animal.

The challenge for animal husbandry both in the past and today is to see beyond the surface. If one could see inside the animal and into the structure of its flesh and intramuscular fat or "marbling," then it would be easier to make decisions on which animals to mate in order to achieve desired goals. Current ultrasound technology is helping breeders to see the insides of breeding cattle. Yet, to be even more ambitious, if one could see into the future, if one

14. Youatt, *Cattle*, 191. Also see William Youatt's *The Complete Grazier and Farmer's and Cattle-breeder's Assistant: A Compendium of Husbandry* (London: Lockwood and Co., 1864).
15. Darwin, *The Variation of Animals and Plants under Domestication*, 178.

could not just imagine but also produce images of projected progeny from a particular dam and sire pairing, then animal husbandry would have taken a revolutionary turn. It is just such a turn that is underway today as technology has created new means of breeding through artificial insemination (known in agriculture as AI) and genetic selection. While I will discuss both technologies, it is worth focusing primarily on genetic selection since it remains in its infancy but promises the most intensive predictive power and chance both to look below the surface of the animal and to project and alter future progeny.

Artificial insemination technologies were first available in the 1940s and commercially became viable over the subsequent decade. As more efficient mediums were developed for housing semen and better freezing methods were developed for storage and transportation, AI allowed a few bulls with valued traits to inseminate thousands of cows. In cattle breeding the male's traits dominate with nearly 90 percent of the herd's characteristics coming from the sire over the last three generations. Consequently, being able to transport bull semen across the country, and even globally, has profound effects. The US National Research Council's Committee on Managing Global Genetic Resources explains that

> technology advances for storing and transporting semen and embryos have allowed the trend toward widespread use of germplasm from a small number of individuals which could cause overall genetic diversity to decline. Regardless of geographic location, when production and marketing systems become more uniform or when the influx of foreign germplasm of one breed is allowed to dominate the population, the danger of narrowing the genetic base of livestock species exists.[16]

Technologies today allow semen companies to "stretch" bull sperm over 20 times farther than early 1950s AI. With such improvements, a few bulls are able to extend their influence on the breed

16. Committee on Managing Global Genetic Resources, *Managing Global Genetic Resources: Livestock* (Washington, DC.: National Academy Press, 1993), 6.

as a whole. Before addressing genetics directly, it is worth simply noting that AI produces a new breeding economy. While choice bulls have always been a prized commodity, until AI, transportation as well as limits to a bull's number of ejaculations confined the influence of any one animal on the species. With artificial insemination the value of these few animals and their impact on the many has increased exponentially. Through transportation of semen from top bulls, the global economy has expanded its pursuit of the singular ideal within a breed. Conversely the possibility for a plurality of ideal types has been lost since global dissemination of select germplasm reduces the number of actual animals introducing their traits into the global gene pool.

The selection of choice bulls for studding is determined by their physical characteristics; however, today sizing up an animal is increasingly *not* done through directly reading the animal's body but rather indirectly through reading grafts and charts which provide another means of accessing the "insides" of the animal. Stated most succinctly, data replaces portraiture as a means of visualizing the animal. The primary means of choosing a bull is its statistical average offspring as compared to other bulls of the same breed. The average weight is expressed in EPDs or estimated progeny differences, also called expected progeny differences. EPDs are a variation on the cattle herdbooks begun with the shorthorn in 1822. Like the herdbook, EPDs were devised as a means of predicting future offspring as well as an attempt to see into the stud bull, to see beyond the visible surface.[17] While farmers have collected information on individual herds for some two hundred years, EPDs provide national (and even worldwide) information about how particular characteristics of an animal translate into future progeny. Data is collected on the birth weight, weaning, and yearling weights of the animal's offspring. This data is compressed to figure an ideal "average" offspring produced by this bull. This ideal

17. For a history of the development of EPDs in the US since the first beef cattle national sire summary in 1971, see John Evans and David S. Buchanan's "Expected Progeny Differences: Part 1, Background on Breeding Value Differences," Oklahoma Cooperative Extension Facts Sheet (Oklahoma State University Division of Agricultural Sciences and Natural Resources), accessed April 19, 2011, http://osuextra.okstate.edu/pdfs/F-3159web.pdf.

is then compared to the "average" offspring of other bulls of the same breed. In EPDs the assumption is that the dame is an average cow of the breed stock. So, for example, the bull New Design 878 has EPDs of +1.5 birth weight, +43 weaning weight, +84 yearling weight. His progeny have pound weights above average in each of the key test categories. With consistently above average progeny, the semen company ABS Global touts their bull New Design 878 as "one of the Angus breed's all-time fastest rising stars. Today, as a three-year-old he ranks as one of the most popular bulls in the industry due to his exciting pedigree, impressive phenotype and balanced EPD package."[18] For his superior offspring, ABS has "awarded" New Design 878 with a branding designation as "RockSolid.™"

While I have been discussing how expected progeny differences track particular "characteristics" of a bull, ABS follows agricultural scientists in referring to EPDs as genetic traits. In fact ABS has trademarked the slogan "Pasture to Plate Genetics" with which it promotes its goods. EPDs are designed to make genetic material of the animal visible through statistical averages gleaned from actual animals. They represent the current hybrid state of affairs in animal agriculture where the science and industry negotiate between age old phenotypic markers and relatively new and increasingly novel genotypic data. The data from actual animals is clearly phenotypical information; however, as it gets compiled into statistical averages, it is said to indicate genotypical traits of the sire.

Genomics is currently a buzzword used for a wide variety of information in agriculture, not all of which is strictly genetic. With EPDs one is not looking at actual genes nor gene markers. Instead, genetics becomes a loose term for the transformation of genes into physical attributes of an animal's offspring which "expresses" these genetic characteristics. In yet another transformation, the physical characteristics of offspring become quantifiable sums that are averaged as data in EPDs. For many cattle breeders, genotype is deduced by phenotype. When semen companies such as ABS advertise the genetic superiority of their animals, they are

18. ABS Global, "Rock Solid from the Start," *Breeders Journal*, Spring (2001): 10.

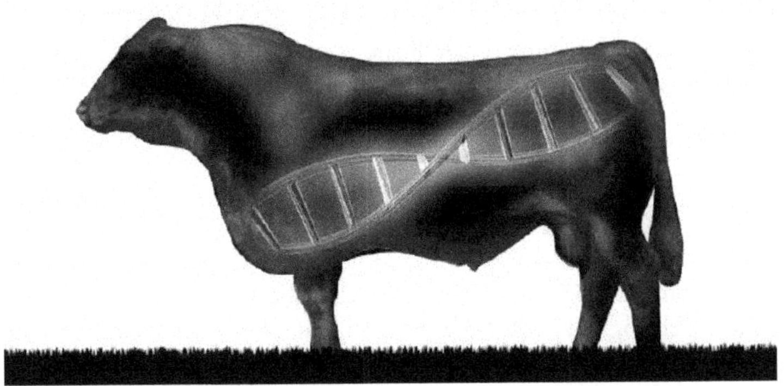

FIG.3. *Igenity-TenderGENE™* advertisement, 2006. Courtesy of Igenity.

not discussing alleles and gene markers; instead, they are referring to statistical averages of physical traits of offspring – qualities that are as much phenotypical as genotypical and that avoid a discussion of molecular biology. Nevertheless, the rhetoric turns cattle breeding away from representing the animal through its surface as ascertained visually or imagined on painters' canvases. The new language of genomics moves representation toward a concept of discovering the "inside" of the animal and depicting its ideal progeny through statistical EPDs.

Nowhere is the idea of knowing the interior of the animal through genetics more evident than in the work of Igenity, a subsidiary of Merial, a leading animal health company. Indeed the very name "Igenity" blends the words "identity" and "gene" so as to place the company at the intersection between the two. Igenity's most recent advertising campaign furthers the claim for genetics as a representational system (fig. 3). The ad shows a traditional side view of an Angus steer in silhouette and standing amid a grass field. A colorful depiction of a model DNA double helix stretches across the darkly shaded bovine body. The nucleotide bases adenine, thymine, cytosine, and guanine glimmer in pastel colors as they interlock along the length of the animal. The text for the ad reads "Nothing improves confidence like inside information."[19] "Inside information" as a phrase denotes privileged information from which one can make a profit, as in "insider trading" on the stock market. Yet in the advertisement, "inside information" is most obviously depicting knowledge of the animal through a knowledge of its genetics. To know its genes is to know its insides.

Knowing the insides of beef cattle means knowing the flesh as meat for consumption. Igenity offers several genetic tests to determine the quality of meat of a particular animal. Cattlemen simply mail to the company several hair follicles from the animal they want tested. Igenity-L™, for example, tests for the leptin protein. Leptin is a hormone that plays a role in metabolizing fat and thus affects the "marbling" of beef. DNA variation affects an amino acid which determines the folding of the hormone. Animals with a higher "t" allele grade yield higher quality USDA meat while the c allele gives a lower quality meat. Essentially, the Igenity-L™ test determines the t allele frequency in relation to the c allele. As Igenity explains it "in an average group, animals that look identical when they are sorted will finish differently. Some of this difference is because of different leptin genotypes."[20] Looks can be deceiving but genetics provides an "insider" knowledge into the material construction of meat and fat. In another product, the Igenity-TenderGENE™ test yields data on markers associated with the calpain and calpastatin genes. The company tests at two SNP in the calpain gene (SNP 316 and SNP 4751), and at one SNP in the calpastatin gene (SNP UoG-CAST1). If these markers have high cytosine bases, animal flesh is more likely to be tender as determined by the Warner-Bratzler Shear Force (WBSF), a standard for the industry. Igenity of-

19. I would like to thank Igenity for granting use of this image without charge. Interestingly, this advertisement image is not run by the UK branch of the company since viewers there associate the image with genetic modification. While all selective breeding is to some degree human modification of genetic traits in animals, Igenity is not a "gm" company in the common use of this word.

20. Igenity, http://us.igenity.com/igenity_beef3.html. Note that since this essay has come to publication, the company has expanded and changed its website. Much of this information and more is still available but with different wording and URLs.

fers several other genetic tests including a test for color (important since Angus beef is currently defined as any beef cattle with a hide that is 51% black), a parent matching test to determine the sire in a multiple sire commercial ranch, and a DNA trace that acts as a genetic "fingerprint" for the animal and can serve as a tracking device as the cattle moves through multiple ranches. While not advocated by many cattlemen, such DNA tracking devices could prove useful during the breakout of a disease.

Igenity extends their trademark from the tests to the actual cattle genes themselves. The company owns several gene markers for which it tests. Those that it does not own, it leases from other companies or individuals who have trademark rights. Royalties on these trademarked genes ensure that the owners of these genes can decide the direction of research and breeding. The notion that cattle are property extends to the very basic building blocks of the animal and to the "depths" of its "insides." No longer do we brand the animal hide with a Circle K or other such ranch names; instead the branding takes place at microlevels and with a ™ and patent rights. Here we have reached the furthest frontiers of colonizing animals by planting our flag on the most elemental means of self-replication.[21]

While branding was something done to the hide, genetic branding has become the new terrain for marketing animals. The US Department of Agriculture (FDA) grades meat as choice, select, or prime based solely on intermuscular fat. In order to realize greater profits for high-end beef and to create a homogenous and recognizable product, beef companies are fashioning niche markets and unique brands beyond the USDA designations. Most cattlemen will tell you that they are trying to do with cattle what Frank Purdue has done with chickens – create a homogonous and recognizable taste and attach such qualities to a particular label.[22] In part, this new branding is done through genetics. Angus cattlemen have had notable success in marketing various brands of Angus meat with 36 different programs currently monitored by the FDA. Each brand touts the genetic qualities of its cattle, for example: Certified

21. For an extended discussion of corporations owning animal genetics, see Donna Haraway's reading of OncoMouse™ in her *Modest Witness@ Second Millenium* (New York: Routledge 1997), 78-85.
22. Michelle Conlin, "Brand that cow," *Forbes*, 163:8 (April 19, 1999): 76.

Angus Beef (CAB) advertises to ranchers by telling stories about cattlemen who have used "Angus genetics" to achieve CAB status, ABS has partnered with Premium Gold Angus and they "are looking for cattle with known Angus genetics" to add to their supply chain while National beef has a Farmland Black Angus brand with "genetically consistent information." Currently I know of no program that actually tests DNA of cattle to determine if the animal can be certified. Instead, "genetics" means phenotypical traits as well as marbling and tenderness based on EPDs and actual data from the slaughterhouse. Nevertheless, actual genetic tests may not be far off in order to better distinguish particular brands. In this respect Igenity with its Igenity-TenderGENE™ test may be forecasting the future of breeding branded cattle for branded meats.

Cloning provides yet another means of knowing the offspring of an animal. Somatic cloning entails taking the donor cell of the animal to be cloned and placing it within the hollowed out or enucleated oocyte. After nuclear transfer the cell is fused and allowed to grow in vitro until it forms a blastocyst, at which point it is transferred to a recipient cow.[23] The cloned animal is born at the completion of gestation. The process itself is expensive, roughly $20,000, which makes it an inefficient method for creating more beef. However, in the case of prize bulls whose sperm is valued, one or several clones could increase the amount of sperm produced by the bull and his "copies" with the resulting semen fees after only one season more than making up for the cost of cloning. The clone itself is not eaten but rather consumers buy the meat from the offspring of a clone; at least in theory, since very few cloned animals' meat or its offsprings' have found their way to market yet. In November 2005, the Veterinary Medicine Advisory Committee for the FDA found no substantial differences between the meat of cloned and non-cloned animals. Despite a "yuck" factor from consumers, in 2006 the FDA announced that clones and offsprings of cloned animals are safe to eat. The public is comfortable with a

23. For a link between human and cattle reproductive technologies see Amade M'Charek and Grietje Keller, "Parenthood and Kinship in IVF for Humans and Animals" in *Bits of Life: Feminism and the New Cultures of Media and Technoscience*, ed. Anneke Smelik and Nina Lykke (Seattle: University of Washington Press, 2008), 61-68.

wide variety of other genetic modifications of animals including AI as a prosthetic for actual animal sex and genetic testing as a tool for knowing animal flesh. Such technologies are not considered to be changing the animal despite the degree of modification they enable. In contrast, consumers seem to be uneasy with cloning because they perceive the technology as too invasive and manipulative.

Branding, cloning, and genetics provide new means of targeting particular characteristics of cattle and making demands upon the animal to "beef up" these traits. While ranchers work to know and modify the inside or interior of the animal, what remains under question is the animal's interiority. Do cattle have a "world view," a *gestalt*, and how might genetic modification through breeding effect this view from within?[24] Or put more forcefully, albeit playfully, do cattle with increased tenderness experience the world differently from their ancestors? To a great degree, this is an unanswerable question, or as Nagel says of bats:

> In so far as I can imagine this [what it is like to be a bat] (which is not very far), it tells me only what it would be like for me to behave as a bat behaves. But that is not the question. I want to know what it is like for a bat to be a bat. Yet if I try to imagine this, I am restricted to the resources of my own mind, and those resources are inadequate to the task.[25]

Yet, the very futility of asking what it is like to experience the world as cattle do is the usefulness of this question, since it requires us to create a new means of evaluating animals. This question allows for a particular interiority of the animal and precisely to the degree that we can not access it, such interiority remains just that, interior and not externalized. Essentially, such an interiority, evades representation and the accompanying technological refashioning of the

24. For a discussion of animal gestalt see Jakob von Uexküll's "A stroll through the world of animals and men: a picture book of invisible worlds," in *Instinctive Behavior: the development of a modern concept*, ed. Claire H. Schiller, trans. D. J. Kuenen (New York: International Universities Press, 1957), 5-80.
25. Thomas Nagel, "What is it Like to be a Bat?," *The Philosophical Review* 83:4 (October 1974): 435-50.

interior through breeding. It is, one hopes, an interiority without end, by which I mean not simply that cattle will always have an interiority but that such interiority remains, remaining without the teleological and evolutionary ends we have put to cattle, fashioning them from birth to their end as meat (and even before their birth in the selection of sire and dam). Our inability to allow such interiority in the West does not preclude its possibility. That is to say, the interior of the animal does not have to be exposed as something representable, knowable, and consumable. In our management of the animal interior, we subsume the animal's interiority as well. Yet, certainly one can think of other times and places where cattle have a place in relation to humans but without an absolute appropriation by human knowledge and commoditization. Hindu and Buddhist treatment of cattle begin to suggest such a possibility.

The cattle problem, then, is one of interiorization. To write on the interiorization of the other, the animal other, will have to be another essay for and from a different time. However, I would like to sketch out parts of it here as the end of animals as meat meets the possibility of an interiority without end. Certainly, an initial step would be to explore the economy of evolution in which genetic traits circulate, and to place this in relation to the economy of breeding from Robert Bakewell to Igenity. In *Acquiring Genomes*, Lynn Margulies and Dorian Sagan develop a theory of evolution based on ingestion. While "acquiring" seems to suggest capitalist ventures in trading genomes across the globe, these microbiologists are actually interested in a much older story of evolutionary change.[26] They observe how species gain a series of genetic markers through failed digestion. Remnants of the species that has been eaten live on in the host agent as genetic traces. These newly acquired genes help the host change and adapt to its environment.

Moving from genetics to anthropology, human social transformation by eating beef is itself a long story; as cattle and their meat make changes possible within human social groups, the animal as (in)digestion remains with us. Consequently, cattle form a

26. This passage is from a transcript of Donna Haraway's keynote address at The Society for Literature, Science, and the Arts conference held a Duke University in 2004. The address aligns with her contribution in a special issue on animals in *Configurations* 14:1-2 (2006), 97-114.

particularly odd "companion species" in Haraway's terms since, unlike the dogs that she writes about in *The Companion Species Manifesto*, we eat our companion cattle. This then leads to the question of the subject and as Derrida asks:

> [t]he question is no longer one of knowing if it is "good" to eat the other or if the other is "good" to eat, nor of knowing which other. One eats him regardless and lets oneself be eaten by him. . . . The moral question is thus not, nor has it ever been: should one eat or not, eat this and not that . . . but since *one must* eat in any case and since it is and tastes good to eat, and since there's no other definition of the good (*du bien*), *how* for goodness sake should one *eat well* (*bien manger*)?[27]

In eating this other, what remains? And, as asked above, is there a remainder that is for the cattle unto themselves? What might be the relationship between this remainder and indigestion? Given the very high percent of Americans and Europeans who are overweight, the problems of incorporation of cattle can even lead to issues of morphological change of humans in the 21st century. That is to say, we manage cattle and we eat them, manage and *manger*. Because we eat them we shape and manage them a particular way. Incorporation as eating is managed by corporations as businesses. Corporate chains from fast food icons such as McDonalds to steak houses such as Longhorn (which serves more polled than horned cattle) make demands on breeders and farmers. So, we change the beast according to these demands.[28] Yet, they change us as we digest globally homogenized and overly processed animal products.

Within the scientific and agricultural practice of cattle breeding it is worth noting the lack of conversation about the animal-in-itself.[29] The Committee on Managing Global Genetic Resources

27. Jacques Derrida, "'Eating Well' or the Calculation of the Subject," in *Who Comes After the Subject*, ed. Eduardo Cadava, Peter Connor, and Jean-Luc Nancy (New York: Routledge, 1991), 114-15.
28. Michael Pollan, *The Omnivore's Dilemma: A Natural History in Four Meals* (New York: Penguin Press, 2006).
29. P. M. Visscher, et al., "A viable herd of genetically uniform cattle," *Nature* 409 (18 January 2001): 303.

have reported their concern over the reduction of genetic difference within a species through modern breeding practices; however, this has done nothing to effect the economy of industrial breeding. Instead, it merely reserves a place for genetic difference in databanks and cryogenic sperm banks. Concerns for any "interiority" of the animal are also noticeably absent from the US National Research Council report on the health issues related to biotechnological farm and research animals.[30] At most, scientists discuss how technologies such as embryo transfer and cloning cause mutations resulting in deformity and, at times, early death. While the suffering of individual animals is certainly of importance and still not sufficiently emphasized in many studies, what is lacking here and elsewhere in scientific discourse is a conversation about breed and species "mutation" which takes place through selective breeding. "After Dolly: Ethical limits to the use of biotechnology" in *Theriogenology*, a leading journal in animal reproduction, states the case quite plainly:

> Often they [concerns regarding genetic manipulation of animals] are expressed in terms of boarders and limits. They indicate that to some people there are as it were freestanding limits to the degree of control that humans can ethically exercise over other living beings. Concepts such as integrity and dignity are often evoked to describe these limits. There is a tendency in the [scientific] academic literature to write off such limits as irrational and/or religious, but lack of conceptual clarity should not lead to dismiss them too quickly.[31]

Unfortunately the authors do not address these concerns but rather state them as a problem of public opinion to be corrected or negotiated by scientists. At stake, however are larger questions outside the borders of typical scientific discourse, questions which address the role of technology and anthropological questions about the role of human behavior for "companion species." Knowing the

30. Debra Davis ed., *Animal Biotechnology: Science-Based Concerns* (Washington, D.C.: National Academies Press, 2002).
31. Jesper Lassen and Mickey Gjerris and Peter Sandøe, "After Dolly: Ethical limits to the use of biotechnology," *Theriogenology* 65 (2006), 999.

animal's insides is not knowing its interiority or "what it is like" to be beef cattle. Genetics has extended and is poised to supercede phenotypical methods of breeding and shaping animals inside and out, in what appears to be a totalizing knowledge; this data that proposes "insider knowledge" in fact flattens the animal by neglecting the folds of its interiority, the possibility that there is an animality of the animal outside the horizons of the human.

This is not to say that the interiority of the animal, the animal-in-itself, is more natural than our uses of domesticated breeds. Rather, the problem should be posed differently: what are the environments in which humans and animals dwell, where do these environments meet and where do they diverge. By rephrasing the question, it is possible to avoid the problematic dichotomy of nature-culture. The animal-in-itself should not be a nostalgic story of animal "origins" or edenic beginnings. This is where backbreeding cattle to achieve the auroch fails to make sense. Backbreeding becomes another human constructed teleology for the animal. Margulies and Sagan provide a different sort of story in which there has always been digestion and indigestion. This eating takes place within a context, a forgotten environmental background. The animal-in-itself points to these neglected worlds, the "what it is like" that remains inaccessible. Our current uses of genetics in breeding animals considers the human economies without regard for larger and other domains.

Networked Encounters: Presence, Pattern, and the Original Body

Maria Chatzichristodoulou (aka Maria X)

> The term one doesn't hear so much any more is "original".
> ... we've begun to wonder what it might mean.[1]

In this paper I address notions of presence and absence in networked encounters. In the first instance I am using the notions of presence and absence to describe the condition and experience of situating (for presence) or excluding (for absence) one's corporeal body and "aura"[2] within /from a specific spatio-temporal context, and a set of relationalities that include (the) "other"(s). Furthermore, I employ the term "networked encounters" to refer to (performance or performative) encounters that are, to some degree, network-enabled. Such encounters depend on interconnectivity, which makes them porous, penetrable, and thus constantly "open" to new inputs and – unexpected – change.

My project here is to suggest that the conceptual dichotomy of presence-absence is not appropriate for an analysis of the hybrid states-of-being that a cyborgian, posthuman creature[3] finds her/himself in. I argue that concepts of presence and absence depend on a series of polar opposites, central in which are the binaries of materiality versus immateriality, and originality versus artificiality. More specifically, I argue that the notion of presence is bound with the assumption that the corporeal body is an "original," whereas all its representations, reproductions and/or extentions are "artificial." I suggest that the materiality of the physical,

1. Thomas McEvilley, "Marginalia," in *Origins Originality + Beyond* (Sydney: The Biennale of Sydney, 1986), 30.
2. "Aura" is, according to Walter Benjamin: "A strange weave of space and time: the unique appearance or semblance of distance...". See Walter Benjamin, "A Small History of Photography," in *One-Way Street and Other Writings,* trans. Edmund Jephcott & Kingsley Shorter (London: NLB, 1979), 250.
3. According to Sarah Kember, the terms cyborg and posthuman are not synonymous since, although they share similar ontologies and epistemologies, they do not necessarily share the same politics, history or ethics. See *Cyberfeminism and Artificial Life* (London & New York: Routledge, 2003), vii.

carnal body is what "safeguards" its corpo-reality,[4] and assumed "originality." I then go on to argue that ideas of originality and artificiality in relation to the body are not pertinent within a posthuman context. Furthermore, I present and elaborate on N. Katherine Hayles's proposal for the introduction of a new dialectic of pattern-randomness as complementary to the existing one of presence-absence.[5] Hayles argues that the pattern-randomness dialectic could provide a useful framework for the analysis of encounters that occur between hybrid bodies made of information and flesh.

Finally, I offer a case-study through Entropy8Zuper!'s performance/software/net art piece *Wirefire*, 1999-2003. I believe that this work exemplifies my argument as it explores the experience of bodily desire in virtual, networked encounters, where physical presence and absence are translated into informational pattern and randomness. As an audience/ participant of *Wirefire* during the last two years of its performance I used to ask myself: What becomes of my corporeal body and aura once I enter this virtual performance space? Where is desire located within this context? How is it constructed and experienced? How do the networked *Wirefire* encounters affect my embodied self?

A *Wirefire* Encounter or the Presence-Absence Paradox

> I want to see how people without bodies make love.[6]

I am at home. It's Thursday night, about one AM Athens' time (Greece). I sit on the sofa, holding my laptop. I log on to Entropy8Zuper!'s website[7] to attend a net performance piece,

4. The term emphasizes the quality of the "real" in relation to the "corpus," that is, the carnal aspect of the body. In my view, the term corporeal hints that representations, reproductions and extensions of the body that are not carnal, are also not "real."
5. See N. Katherine Hayles, *How We Became Posthuman: Virtual Bodies in Cybernetics, Literature, and Informatics* (Chicago and London: University of Chicago Press, 1999).
6. Allucquère Rosanne Stone, *The War of Desire and Technology at the Close of the Mechanical Age* (Cambridge, MA. and London: MIT Press, 1995), 38.
7. http://www.entropy8zuper.org (accessed April 19, 2011).

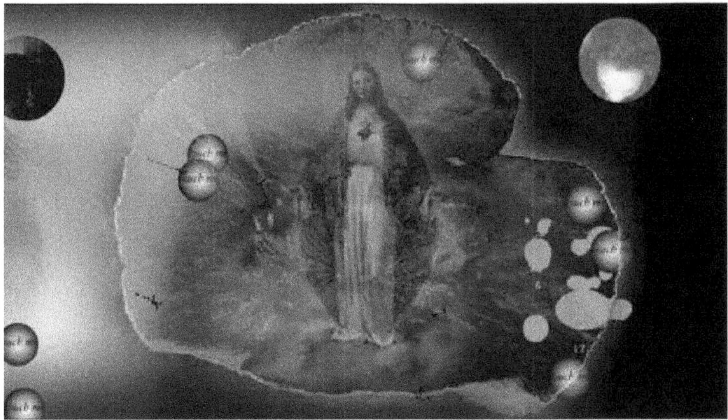

FIG. 1. *Wirefire.*

Wirefire.[8] *Wirefire* is all about sex: the story behind each performance is about a couple, dispersed – she in New York City, he in Ronse (Belgium)[9] – who meet once per week to make love online, for and with audiences. Entropy8 (Auriea Harvey) and Zuper!'s (Michael Samyn) act of online, virtual love-making is no porn-stream – no photorealistic representations of sexual encounters are available on this site. What I witness is an act of love translated into visual poetry: Auriea and Michael mix still images, flash movies, animations, sounds, live streams and text files in real time to create "compelling and seductive narratives"[10] that evolve through visual and sonar landscapes, all of which narrate the same story: "being digital and being in love."[11]

While I'm logged on the website I can just "lurk" and watch the live performance unfold: images, movies, texts, and sounds, are

8. http://www.e8z.org/wirefire (accessed April 19, 2011). Wirefire started in 1999 and lasted for four years, till 2003. Documentation of the piece is available on Entropy8Zuper!'s website.
9. This is how the performance started. Later Auriea moved to Belgium to live with Michael, which means that most of the *Wirefire* performances were actually performed by both artists situated in the same physical space.
10. According to the SFMOMA Webby Prize 2000 Jury comprised, among others, by Machiko Kusahara, Gary Hill, and Benjamin Weil. *Wirefire* won the prize. See http://www.sfmoma.org/press/press/press_webby.html (accessed March 2004).
11. http://www.entropy8zuper.org (retrieved March 2004).

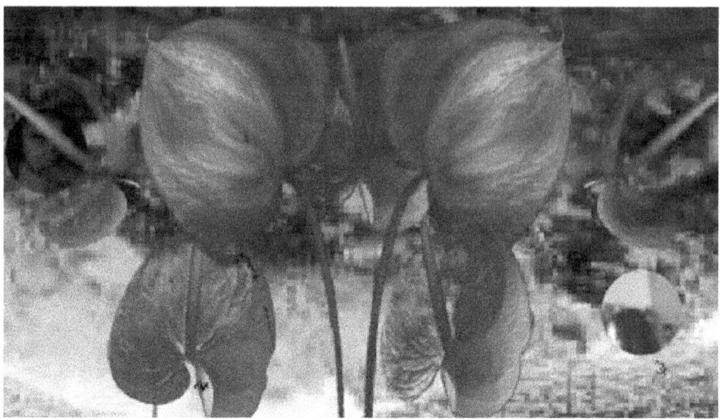

FIG. 2. *Wirefire.*

mixed live from E8Z! into a magnificent, "baroque"[12] composition. Live video streams of the authors appear in bubbles and integrate into the audiovisual mix. Abstract, fragmented narratives emerge: E8Z!'s performance talks about romance, love, desire, passion, eroticism, and sex. It also talks about broader notions of love and hatred[13]: E8Z! use symbols such as flocks of sheep and bees, fire, the Bible, to question the way conditions of love and hatred are created on a larger scale as global phenomena. *Wirefire* brings forth issues of religion, fanaticism, violence, war-but also peace, tranquility, physical and spiritual love, beauty, the (futile?) search for paradise... While I watch, the artists and other audiences can see that someone is watching: every audience member that logs on the website appears on everybody else's screens as a spec of dust. That is all they know about me though: that I am, somehow, "present." And that is all I know about other audiences: I can find out how many people are watching by counting the specs of dust on my screen, but I don't know who these people are, what they look like, or where they are physically located... I know nothing about

12. Along the lines of Anna Munster's comparative analysis of baroque and digital art. Munster argues that "Both baroque and digital spaces engage the viewer visually, seductively and affectively." See *Materializing New Media: Embodiment in Information Aesthetics* (Hanover, New Hampshire: Dartmouth College Press, 2006), 6.
13. E8Z!, private email communication, 2006.

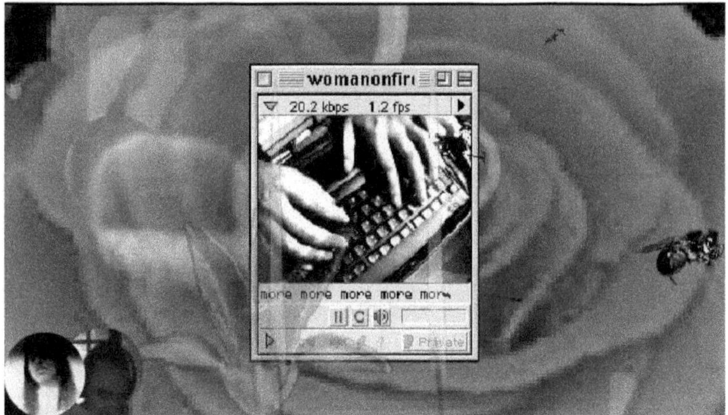

FIG. 3. *Wirefire.*

them other than the fact that they are, somehow, "there," on/in the screen, and we are watching this performance together.

If I want, I can expose myself a little further by contributing my own fantasies, and every other audience member can do the same: *Wirefire* audiences can contribute text and live streams that E8Z! mix together with their own files from the *Wirefire* database, and integrate within the audiovisual landscapes. These are anonymous contributions, and the artists have complete control over whether, when, and how these audience inputs appear on screen. I type in "Be with me." The text appears on the bottom of the page as an anonymous chat input, along with lines other audiences have typed. In the meantime, the performance is going on: brightly coloured sunflowers burst open across the screen. I type in "Feelings float." This never appears as the performance has moved on – there is no text on screen now, only the flowers that have spread across and covered the screen on the sound of an operatic voice. By now, I'm not just sitting on the sofa – I'm also lying in a field of sunflowers, floating aboard the wings of a bird, sitting in front of a raw steak, or burning in hell. Although my corporeal body is always in Athens, I am participating in Entropy8Zuper!'s performance of love-making. My aura is stretching across two spaces, one physical, one virtual: I find myself in a hybrid space that is a mixture of reality and fiction, involved in intimate encounters

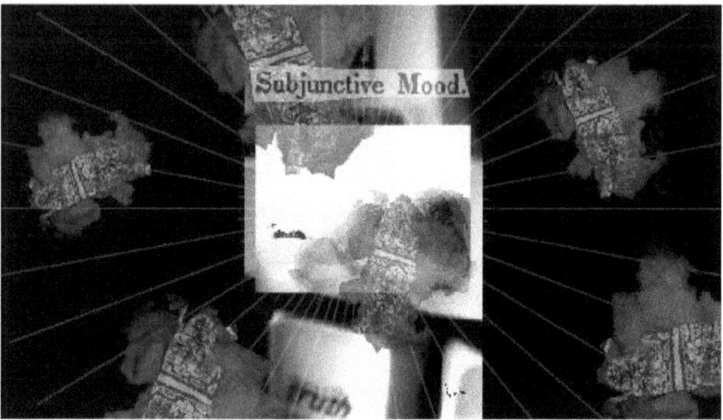

FIG. 4. *Wirefire.*

with strangers who are present on /in (?)[14] my screen…

According to Peggy Phelan, "In performance, the body is metonymic of self, of character, of voice, of "presence""[15]; when it comes to (semi-)mediated, networked, and other forms of technologised performance[16] though, as Hayles points out, "questions about presence and absence do not yield much leverage."[17]

14. See Sherry Turkle, *Life on the Screen: Identity in the Age of the Internet* (London: Phoenix–Orion Books, 1997), 21.
15. Peggy Phelan, *Unmarked: The Politics of Performance* (London and New York: Routledge, 1993), 150. Phelan makes a point about the use of the body as metonymy in performance.
16. I use the term "mediated performance" to refer to genres such as radio and TV drama, and other forms of performance that depend entirely on (a) medium(s) for their actual "being." I add the prefix "semi" in parenthesis so as to include new genres of performance that make use of media but do not depend on them entirely. Many networked performances are semi-mediated since they use networking technologies (e.g. Internet, mobile, telecommunication technologies) as an internal element of their dramaturgy and/or action while, at the same time, provoke immediate encounters among their audiences, between audiences and performers, or even between audiences and their environment (turning this into a stage). An example of semi-mediated performance is British group Blast Theory's works *Can You See Me Now?,* and *Uncle Roy Is All Around You,* where audiences experience both mediated and immediate encounters. For more information, see www.blasttheory.co.uk (accessed April 19, 2011).
17. Hayles, *How We Became Posthuman,* 27. Hayles specifically refers to the technologies of virtual reality rather than any technologised performance practices. Nevertheless, I consider that the problematics these technologies raise around issues of presence and absence apply to

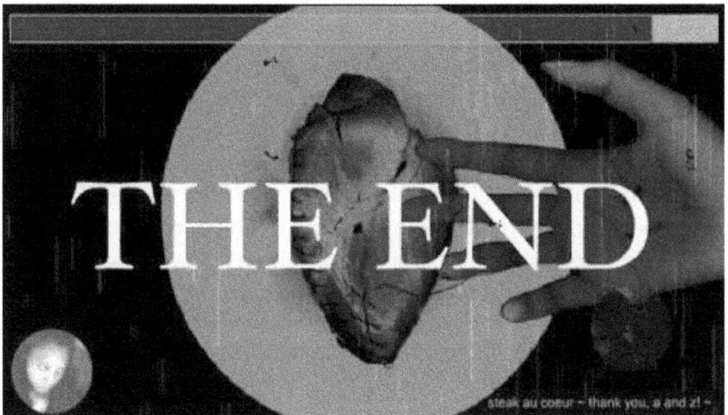

FIG. 5. *Wirefire.*

Phelan's discourse refers to visceral performance staged in physical space, where the performer's corporeal body exists "in the plenitude of its apparent visibility and availability."[18] Hayles, on the other hand, discusses situations where "the avatar both is and is not present, just as the user[19] both is and is not inside the screen."[20] Whereas Sherry Turkle, in her exploration of online communication environments such as MUDs[21] and MOOs[22], wonders whether we have altogether migrated within our computer screens: as we become "increasingly intertwined with technology and with each other via the technology", she asks, "are we living our lives on or in the screen?"[23] Turkle suggests what we have, by now, come to accept: that our enmeshment with technology has made us cyborgs, "transgressive mixtures of biology, technology, and code."[24]

most emergent, hybrid forms of performance that employ telematic, networking and/or other digital technologies.

18. Phelan, *Unmarked*, 150.
19. Hayles employs the term to refer to the user of a VR environment. In techno-performative practices the term "user" could apply to both the performers and the audiences.
20. Hayles, *How We Became Posthuman*, 27.
21. Multi-User Dungeons.
22. MUDs Object Oriented.
23. Turkle, *Life on the Screen*, 21.
24. Turkle, *Life on the Screen*, 21.

In embodied, physical encounters, like the performance encounters Phelan refers to, the notion of presence is clear: present is a person manifested through a physical, living body, that you can perceive with your senses and intellect in proximity to yourself; someone whose body you can see, smell, touch; someone you can talk to and expect a response from in "real time"; someone who is physically tangible within the present moment. In networked and semi-mediated encounters though, as Hayles, Turkle and others[25] have observed, "presence" is not equally straightforward. When it comes to such encounters, the bodily presence, self-evident in its corporeality – the pure, "absolute" presence-as-we-know-it from the physical world (that is, the opposite of absence) – mutates into something else. This new "morph" is no more distinct from – let alone opposed to – the notion of absence. Within this context presence can be perceived as absence and the reverse; presence and absence become interwoven like two sides of the same coin, impossible to disengage from their permanent entanglement: they become a "presence-absence."

This paradoxical state of presence-absence is a hybrid between relative[26] physical absence or dislocation (the user's corporeal body is absent/dislocated in relation to a partner or a physical action, either periodically or throughout the encounter/action), and relational presence (despite her/his physical absence/dislocation the user can still relate to the partner or take part in the action via a medium). I argue that this paradoxical, hybrid state-of-being calls for new approaches to the notions of presence and absence. Once the self exceeds and expands the limits of the corporeal, human body to exist as a cyborg, "a hybrid of machine and organism, a creature of social reality as well as a creature of fiction,"[27] physical proximity/distance no more con-

25. Such as Allucquère Rosanne Stone, Eduardo Kac, Machiko Kusahara, among others.
26. I am referring to "relative physical absence" as I consider that physical presence depends not just on the corporeal body but also on a socially constructed self situated within a specific social and spatio-temporal context, which overlaps with but also exceeds and expands the body.
27. Donna Haraway, "A Cyborg Manifesto: Science, Technology, and Socialist-Feminism in the Late Twentieth Century," in *Simians, Cyborgs and Women: The Reinvention of Nature* (New York: Routledge, 1991), 150.

stitute the criteria for presence and absence. Hence presence and absence can neither be safely identified as conditions nor attributed as qualities any more. Certainties dissolve, as the presence-absence state is based on doubt: presence needs to be manifested in order to be perceived; and it needs to establish the validity of its manifestation(s) for it to be accepted as an "authentic" condition or quality. Even so, it incorporates and coexists with (physical) absence. Absence, on the other hand, also needs to be secured, whereas it can no more be safely assumed to be pure, vacant of any presence.

At this point of ambiguity there is nothing to ensure that what we perceive as presence is not, in fact, an absence; and what we consider as absence does not "leak" traces of presence. I ask: is the notion of presence as doubt, as an ambiguous condition or quality that co-exists with rather than opposes absence something new? Phelan argues that, in his plays, Samuel Beckett "makes clear that presence is doubt; presence is impossible without doubt; doubt is the signature of presence."[28] Moreover, Hayles quotes Roy Walker, who was doing tape-recordings for the BBC in the 1950s, on his experience of "the disconcerting effect ... in the disjunction between voice and presence (he can no longer call himself his own)."[29] Susan Sontag considers a photograph to be "both a pseudo-presence and a token of absence."[30] Finally, André Bazin argues, in his exploration on the ontology of cinema, that "it is no longer as certain as it was that there is no middle stage between presence and absence."[31] Indeed, I suggest that there is a middle (maybe even "central") stage, which was introduced by "the medium,"[32] our extension.[33]

28. Phelan, *Unmarked*, 115.
29. Hayles, *How We Became Posthuman*, 210.
30. Susan Sontag, *On Photography* (London: Allen Lane Penguin Books, 1978), 16.
31. André Bazin, *What is Cinema? Vol. 1*, trans. Hugh Grant (Berkeley, Los Angeles and London: University of California Press, 2005), 97.
32. I consider "the medium" to be anything that mediates information, knowledge or experience.
33. I refer to Marshall McLuhan's influential book *Understanding Media: The Extensions of Man* (Cambridge, MA. and London: The MIT Press, 1994).

The Third Man's Aura

> Aristotle had a great argument called the Third Man .... It goes like this: how can I know that something is an original unless I have another original to compare it to? And how do I know that that one is an original without still another to compare it to? And so on, forever. The original is dissolved in an infinite regress, for the idea of the original is bound to the idea of the finite. .... One function of the idea of the original is to close off a proliferating series, to provide a final term that will keep it finite and manageable, because we need finite series on which to make our projections of meaning and value.[34]

In my view, the presence-absence dialectic depends on and can only be considered in relation to other dialectical oppositions such as: life-death, self-other, white-black, male-female. Such binary systems normally operate on the basis of a privileged term, usually the first part of the binary – for example: presence, life, self, white, male- followed by a second term that is loaded with negative meaning – for example: absence, death, other, black, female. In the context of this essay I will focus on two such discourses which I believe are fundamental in preserving the status of the presence-absence dialectic as dominant today. These are:

a) materiality – as opposed to immateriality

b) originality – as opposed to artificiality and/or simulation.

My question is: how pertinent are these discourses within a postmodern and posthuman context?

a) On Materiality

> Can anyone tell me how to /join #real.life?[35]

First, let us establish the dominance of immateriality as the epistemic discourse of the information era. To that extent, recent

34. McEvilley, "Marginalia," 30.
35. Internet Relay Chat (IRC) enthusiast as quoted by Turkle in *Life on the Screen,* 186.

and current discourses on materiality are seeking to understand how it happened that "information lost its body" and came to be conceptualized as "an entity separate from the material forms in which it is thought to be embedded."[36] This was the outcome of the computer age that brought forth research on artificial intelligence and established the science of cybernetics. Claude Shannon and Norbert Wiener, founders of cybernetics, are generally deemed to be responsible for spreading a disembodied notion of information, conceptualised as distinct from context and matter.[37] Hayles argues that cybernetics initiated the process of the "erasure of embodiment" across a broad field of disciplines, whereas it facilitated the equation between humans and machines, which led to the conception of the cyborg.[38]

Does this mean that discourses on materiality are not relevant within a 21st century context? I think that the opposite is the case: our quest for disembodiment – and, effectively, immortality – through the sciences of cybernetics and AI as well as science fiction and popular culture, has produced what we now perceive as dystopic fantasies from which we need to escape. Escape how? By bringing back into focus embodiment and materiality – this time within a more appropriate, posthuman context. This is Hayles' project in *How we Became Posthuman*: she attempts to foreground embodiment and materiality by making clear that thought depends "on the embodied form enacting it."[39] She goes on to propose that the current historical moment provides us with an opportunity to "put back into the picture the flesh" and "keep disembodiment from being rewritten ... into prevailing concepts of subjectivity."[40]

Apart from Hayles, Mark B Hansen, in his book *New Philosophy for New Media*, approaches embodiment (as lent by recent work in neuroscience) as inseparable from the cognitive activity of the brain.[41] Hansen's project is to embrace the materiality of technol-

36. Hayles, *How We Became Posthuman*, 2.
37. See Norbert Wiener, *Cybernetics or Control and Communication in the Animal and the Machine* (Cambridge, MA: MIT Press, 1991).
38. See Hayles, *How We Became Posthuman*, 1-13.
39. Hayles, *How We Became Posthuman*, xiv.
40. Hayles, *How We Became Posthuman*, 5.
41. See Mark B. Hansen, *New Philosophy for New Media* (Cambridge, MA. and London: The MIT Press, 2004).

ogy that makes it contextual and dependent on specificities of location, space and time, while restoring the link between affectivity and body (of flesh), by foregrounding the body as "coprocessor" of information. Furthermore, Anna Munster, in *Materializing New Media*, undertakes the task of unearthing the relationalities between human matter and technological materialities, while she suggests that we need to treat both as "open-ended propositions ... continually in the process of being made and unmade."[42]

Hayles, Hansen, and Munster, although coming from different backgrounds and following different methodological paradigms[43] attempt (I think successfully) to foreground materiality and embodiment exactly within the current historical context of the information era, postmodernism, and posthumanism. Within this context, they argue, the challenge is not to leave our bodies behind in the quest of a utopian (dystopian?), disembodied future where, by being able to become the information we have created, we will achieve immortality. Instead, as Hayles puts it, the challenge is to remember that "information ... cannot exist apart from the embodiment that brings it into being as a material entity in the world."[44] Munster further suggests that, right now, we need to shift our attention from the tension between information and matter towards what she calls "digitality": "a virtual ecology of bodies, technologies and socialities."[45]

b) On Originality

> Our culture is undergoing a truly drastic change in terms of our physical and psychological relationship with space and other bodies. Digital technology ... has brought us the notion of disembodied presence. We can no longer simply believe what our eyes see and our ears

42. Munster, *Materializing New Media*, 13.
43. Hayles comes from an English literature background, Hansen from philosophy, and Munster is a media theorist. In terms of their different methodological approaches, Hansen aims to construct a new phenomenology for new media; Hayles positions her arguments on a combined analysis of history, theory and fiction; whereas, Munster focuses on the aesthetics and ethics of "digitality."
44. Hayles, *How We Became Posthuman*, 49.
45. Munster, *Materializing New Media*, 155.

> hear. Telerobotics makes it possible to represent oneself in far-away places through a network. But how do others know that the robot is operated by a real person? And how do we know that the robot is representing the world accurately back to us?[46]

As stated in my introduction, I consider that the binary of "originality" versus "artificiality", when applied in relation to the body, is neither productive nor pertinent within a posthuman context. I here argue that the conceptual dichotomy of presence-absence as we know it depends on this very binary. In my view, central in these oppositional dialectics is the assumption that the corpo-real body is an "original," whereas all its representations, reproductions, transformations and/or extensions, are artifacts or "artificial." In that sense, our culture often approaches materiality as the "bearer of truth," and immateriality (or its impression) as a "potential trickster." Kusahara's questions "how do we know," "how do others know" clearly express the confusion that follows the loss of our material tokens of truth. Kusahara talks about telerobotics, where what poses these questions is the distance: the robot becomes, for us (its operators) on the other side of the screen/world, its reproduction in a 2D image – so how do we know that the robot actually exists? And how do other people know that we – and not someone else, somewhere else – are the robot's operators, or that the robot does not follow a set of built-in instructions?

If we move on to virtual worlds, the questions multiply. Let us take *Wirefire* as an example: how do we know that Auriea and Michael are performing live every Thursday night? That the performance is not a random remix of files generated by the *Wirefire* database? How do we know that they are really based in USA and Belgium respectively? That the "live" streams are not prerecorded? That the specs of dust on our computer screens do represent audiences? How do we know that Auriea and Michael actually exist? That they are two and not one or many? That they are humans and

46. Machiko Kusahara, "Presence, Absence, and Knowledge in Telerobotic Art," in *The Robot in the Garden: Telerobotics and Telepistemology in the Age of the Internet*, ed. Ken Goldberg (Cambridge, MA and London: MIT Press, 2000), 200.

not bots? Maybe we don't... According to Turkle, "traditional ideas about identity have been tied to a notion of authenticity that such virtual experiences[47] actively subvert. ... the self is not only decentered but multiplied without limit."[48] In emergent virtual/synthetic environments, this decentralisation and multiplicity does not only apply to the question of identity but also to the subject's actual existence (the question is not so much "who are you" but "what are you" or, "are you"?).

Walter Benjamin, in his influential essay "The Work of Art in the Age of Mechanical Reproduction,"[49] discusses aura in relation to presence. Benjamin argues that aura is tied to the presence of an original (artistic, historical, or natural) object , and its "unique existence"[50] at the specific physical place and temporal framework within which it is situated. Concerning the artistic object Benjamin argues that, although the object itself can be reproduced, thus substituting "a plurality of copies for a unique existence,"[51] its aura cannot. What reproduction does, according to Benjamin, is cause the decay of the aura and, along with it, the decay of the original, unique object or situation. "Contemporary masses," he argues, in their urge to bring things "closer" to them, are "overcoming the uniqueness of every reality by accepting its reproduction."[52] In so doing, they are replacing uniqueness and permanence by transitoriness and reproducibility.

Although Benjamin's analysis focuses on the reproduction of the art object, he also approaches film, and discusses the reproduction of the performer's body on screen. He quotes Pirandello as expressing his concern that the film actor feels: "as if in exile – exiled not only from the stage but also from himself. With a vague sense of discomfort he feels inexplicable emptiness: his body loses its corporeality, it evaporates."[53] Consequently, says Benjamin, the

47. Turkle refers to MUDs and MOOs.
48. Turkle, *Life on the Screen*, 185.
49. See Walter Benjamin, "The Work of Art in the Age of Mechanical Reproduction," in *Illuminations*, ed. Hannah Arendt, trans. Harry Zorn (London: Pimlico, 1999).
50. Ibid., 214.
51. Ibid., 215.
52. Ibid., 217.
53. Pirandello in Benjamin, "The Work of Art in the Age of Mechanical Reproduction," 222-223.

aura that envelopes the actor as well as the character s/he incarnates, vanishes along with the corporeal body: "for the first time – and this is the effect of the film – man has to operate with his whole living person, yet forgoing his aura. For aura is tied to his presence; there can be no replica of it."[54] But if aura is tied to the performer's presence, the question I pose is: what constitutes "the performer's presence"?

In Benjamin's analysis presence appears to be tied to the one-and-only, "original," corporeal human body, and thus to be situated within a specific, unique and non-reproducible framework of space-time. Since Benjamin considers presence as unique and impossible to reproduce, and aura as tied to the presence, it follows that aura is also non-reproducible. If we accept Benjamin's argumentation we can conclude that a) all reproductions, representations, transformation and/or extensions of the corporeal body are artifacts or "artificial" (as opposed to "original"), since they are often situated within hybrid space-times and transform the corporeal, human body to a cyborgian, posthuman one; and b) presence cannot be experienced through or attributed to such reproductions, representations, transformations and/or extensions, since presence is tied to the "original," corporeal body. This discussion suggests that what can be experienced through such manifestations of the body /self is either absence (the opposite of presence), or nothingness (as absence still relates to physicality, albeit through an oppositional dialectic). Moreover, it suggests that, since there is only one "original" body, all hybrid bodily manifestations can only be conceived as artificial versions of the one-and-only, auratic, physically mapped, corporeal human body.

In my view, Phelan is approaching the issue of presence along the same lines when she argues that the (performing) body is metonymic of "presence" and self.[55] I ask: which (performing) body is Phelan talking about? Whose presence? And whose self? Phelan, like Benjamin, assumes the presence of the one, auratic, "original" human body in the specific space-time frame of the theatre stage. This body she considers as a reassurance not only of the perform-

54. Benjamin, "The Work of Art in the Age of Mechanical Reproduction," 223.
55. See Phelan, *Unmarked*, 150.

er's presence, but also of her/his actual existence, both in relative (in relation to us, the audiences) and absolute (in relation to life and death) terms. Being fully visible and available, this body, according to Phelan, promises the "lack of Being" as the performer disappears to represent someone/something else.[56]

Nevertheless, when it comes to networked encounters and performances, there are no human bodies and no auratic originalities to be assumed. Networked encounters take place in hybrid space-times that are made, more often than not, through a layering of physical and digital, actual and fictional, natural and artificial elements. Human bodies do not have a place in such environments. Human beings, in order to expand their existence in such hybrid multiplicities of space-time, turn into cyborgian, posthuman creatures, enmeshed "at all levels of materiality and metaphor with information, communication and biotechnologies, and with other non-human actors"; they become hybrids of "organic and inorganic forms and processes."[57] The emergence of such hybrid space-times along with the creatures that inhabit them – us, already posthumans[58] – introduces a "pov"[59] of the world where notions of the "original" or "natural," and "simulation" or "artificial" do not exist in opposition, but in fusion. Within this context, Anders Michelsen calls for a study of the "artificial":

> as something not primarily divided by the inherited dichotomies of man-machine, mechanism-organism, etc., but rather imagined, thus created, from the standpoint of an organizational novum with an ontological contingency beyond inherited determinations and constraints.[60]

56. Ibid., 150-151.
57. Kember, *Cyberfeminism and Artificial Life*, vii.
58. Hayles argues that "the posthuman" is a historically specific construction, which we already have become. See *How We Became Posthuman.*
59. "Pov" stands for "point of view" in virtual environments where the user's presence is manifested through an avatar.
60. Anders Michelsen, "The Imaginary of the Artificial: Automata, Models, Machinics – On Promiscuous Modeling as Precondition for Poststructuralist Ontology," in *New Media, Old Media: A History and Theory Reader*, ed. Wendy Hui Kyong Chun & Thomas Keenan (New York and London: Routledge, 2006), 235.

Michelsen thus encourages us to view the "artificial" as something that is not dependent on an "original" through a set of inherited, oppositional dialectics. This imagined artificial cannot be described as non-original or non-authentic, since it does not copy or otherwise duplicate any original. Instead, it exists on its own accord, its ontology ambiguous, unpredefined and unexpected. Michelsen's call to look at the artificial "from the standpoint of an organizational novum" makes me think of it as yet another "original": what else is not defined by "inherited determinations and constraints" – that is, these of the "original" in relation to which the "artificial" can only exist?

Pattern, Randomness, and "A Romance in the Wires"[61]

what was being sent back and forth over the wires wasn't just information, it was bodies.[62]

The first assumption of a posthuman "pov," as articulated by Hayles, is "the privileging of informational pattern over material instantiation."[63] Hayles argues that information is pattern rather than presence. This does not entail that non-information is the absence of pattern, that is, randomness: scientific developments[64] have shown that information can be identified, paradoxically, with both pattern and randomness. Pattern-randomness do not form a binary opposition, and thus do not follow the same set of oppositional strategies as the binaries of presence-absence, materiality-immateriality, and originality-artificiality; instead, they are bound together in a dialectic that makes them complementary to one another.[65] Such systems are not comprised by one privileged and one negative term: randomness is not seen as the absence of pattern, but as the ground for pattern to emerge. "If pattern is the realization of a certain set of possibilities," explains Hayles, "randomness is that much, much larger set of everything else."[66] Since pattern

61. E8Z!, private email communication, 2006.
62. Stone, *The War of Desire and Technology*, 7. Stone refers to phone sex.
63. See Hayles, *How We Became Posthuman*, 2.
64. Such as the chaos and complexity theories.
65. For a more detailed analysis see Hayles, *How We Became Posthuman*, 25.
66. Hayles, *How We Became Posthuman*, 286.

and randomness are complementary rather than oppositional, we do not need to distinguish between one or the other (as we do with presence-absence): a system can integrate both -indeed it normally does- in variable degrees and combinations.

Hayles goes on to examine how notions of pattern and randomness apply, not just in formal theory, but also in "everyday life" situations such as virtual environments.[67] She concludes that in such environments, the focus shifts from questions of presence and absence – which are impossible to answer in any satisfactory way due to the presence-absence paradox – to questions about pattern and randomness, such as: "What patterns can the user discover through interaction with the system? Where do these patterns fade into randomness? …. When and how does … noise coalesce into pattern?"[68] Hayles suggests that we look at notions of pattern as the outcome of our interactions with the system and other users, as complementary to presence; and at notions of randomness as the outcome of the noise created by stimuli that cannot be encoded within the system, as complementary to absence. Randomness can turn into pattern when extraneous stimuli merge together, whereas pattern can gradually fade into randomness. Hayles claims that pattern-randomness systems, because they have no meaning front-loaded into them, evolve towards an open future marked by unpredictability. On the other hand, presence-absence systems depend on a stable origin, which leads them to evolve towards a known end. In the case of pattern-randomness systems, explains Hayles, meaning is made possible – though not inevitable – by evolution.[69]

As discussed earlier[70], one of the reasons the presence-absence dialectic is not sufficient within a posthuman context is its dependence on notions of originality and artificiality, since presence has been articulated as a quality of the auratic (thus "original") human

67. Hayles primarily looks at VR environments (which are not quite "everyday life" situations yet); nevertheless, I believe that the same or similar issues are raised through other types of virtual or hybrid environments that have become part of our everyday life, such as online multi-user, telematic, and augmented environments.
68. Hayles, *How We Became Posthuman*, 27.
69. See Hayles, *How We Became Posthuman*, 285.
70. See this essay: "The Third Man's Aura, b) On Originality"

body. I argue that pattern-randomness systems are more appropriate for the analysis of networked, posthuman encounters, exactly because they are not based on a coherent origin that would entail a pre-loaded set of meanings. Because they start as free of meaning and oppositional strategies (as a result of their non-stable origin), pattern-randomness systems are also free of any pre-loaded sets of moral judgments. This makes any distinction between "original" and "artificial" meaningless in the context of such systems. As pattern-randomness systems do not recognise origins, they also do not recognise originals or the very notion of originality. This absence of meaning creates a vacant space, an emptiness, which allows for the unforeseen, unfamiliar, and/or novel to occur. The liberating potential of such systems is very useful for the analysis of phenomena that are emergent and still in-flux, such as posthuman bodies, hybrid spacetimes and networked encounters.

I have suggested[71] that changes in material conditions (e.g. hybrid and trans-local environments) and embodied experiences (e.g. decentralisation and multiplication of the self through the use of prosthetics, extensions, and avatars) shift the emphasis toward the notions of pattern and randomness, while pushing presence and absence into the background. Physicality evokes presence, whereas information evokes pattern. Immateriality (as an epistemic discourse) and cyborgian existence (as our current state-of-being) challenge presence, while they endorse information, and to that extent, pattern. Within the posthuman context of the information era established notions of presence and absence cannot but lose their "meaning", as they merge into the asignifying state of presence-absence. Thus, a posthuman creature can be described at any given moment as "more present than absent" or the reverse, which can also be articulated as "producing more pattern than randomness" through her/his connections and engagements[72] with the system and other users. Through these encounters develop patterns, which gradually fade into randomness, which again

71. See this essay: "The Third Man's Aura, a) On Materiality"
72. According to Munster – who quotes Steven Shaviro – connection is to the network (and away from sociality), whereas engagement is an active and ongoing social confrontation. See Munster, *Materializing New Media*, 152.

coalesces into pattern, and so on. Hayles warns us though that this shift from presence-absence to pattern-randomness has had a serious implication, namely, the "devaluation of materiality and embodiment."[73] She repeatedly points out that these two sets of dialectics should not be approached as antagonistic, but as "mutually enhancing and supportive."[74]

Another manifestation of the transition from the presence-absence to the pattern-randomness dialectic is, again according to Hayles, the shift of emphasis from ownership to access.[75] Ownership, says Hayles, implies presence (something must be tangible or identifiable for someone to wish to own it), whereas access implies pattern recognition (as, for example, when someone breaks into a computer system tracing patterns to access information). This transition means a reconfiguration of the private-public distinction: whereas possession (linked to presence) implies the existence of a private life, access (linked to pattern) implies the penetration of information in all private spaces. *Wirefire* is a good example of such reconfiguration, as the performance challenges notions of private and public space by providing access (through observation and participation) to a very "private" moment of E8Z!'s lives: their act of virtual love-making. Auriea and Michael do not perform their love-making as physical bodies/ presences within a private space; instead they turn bodily presence into pattern and randomness created via their engagement with each other, the *Wirefire* technical platform, and their audiences. This "presence" made of pattern and randomness rather than material corpo-reality provides the relational circumstances that allow for audiences to access an intimate encounter, while this remains intimate.

Throughout this essay, I have used E8Z!'s piece *Wirefire*[76] as a case-study in order to illustrate the shift of focus from presence-absence to pattern-randomness, as well as the need for the co-existence of these two dialectics, and their application in the analysis of networked performance and performative encounters. E8Z! describe *Wirefire* as: "sex in a virtual world. The loss of physicality

73. See Hayles, *How We Became Posthuman*, 43-49.
74. Ibid., 48.
75. Ibid., 39.
76. See http://www.e8z/wirefire (accessed April 19, 2011).

-but let me not call it a loss- because of what is gained... What one can gain through *Wirefire* is a new sense of touch, an enhanced fantasy, a glimpse of a personal utopia."[77] *Wirefire* grew out of a very physical necessity: Entropy8, then based in New York, and Zuper!, based in Ronse, met and fell in love on-line. *Wirefire* was created as a result of the couple's need to express their love and desire telematically, and their frustration with the limitations of commercial software. Technically, *Wirefire* is a combination of animations, sounds, images, chat, and live camera streams pushed into motion with an engine of scripts, and built layer upon layer to form interactive improvisational performances. As E8Z! put it, *Wirefire* is their "personal remixer of emotions and graphics": "A thin wire strung between the camera images, fingers on keyboard like fingertips on each other's skin, letters on a monitor screen, whispers."[78]

Wirefire, while active, existed in three different operational modes: Random, Live and Replay. Through the Random mode, the audiences were invited to initiate unique *Wirefire* performances generated by the computer, through a random mix of the files on the *Wirefire* database. In the Live mode, E8Z! performed their love-making live, engaging with each other, the *Wirefire* database, and their audiences in real time, once per week, for four years: every Thursday night at midnight (Belgium time). Their cybersexual encounter was a poetic love-performance-making, abstracted from physicality in terms of both matter and form. Finally, the Replay mode is a documentation of the live performances – albeit not a very accurate one, as the live streams could not be archived for technical reasons.

E8Z! invited audiences to generate random versions of visual and sonar landscapes through the *Wirefire* database, as well as take part in the live *Wirefire* performance. For four years E8Z! performed the same story: being digital and being in love; experiencing desire for a pixel image; exploring what ecstasy is like in the network... For E8Z! and their audiences to perform *Wirefire*, they endlessly reproduced themselves as sound and image data and the force that

77. http://e8z.org/wirefire/SECRET.HISTORY/A_SECRET_HISTORY_OF_WIREFIRE.txt (retrieved March 2004).
78. http://e8z.org/wirefire/SECRET.HISTORY/A_SECRET_HISTORY_OF_WIREFIRE.txt (retrieved March 2004).

pushed these data into motion. Their digital selves became abstract audiovisual "bodyscapes" and their engagements became rhythm, colour, pattern, sound and text. They themselves became acentered and distributed: by networking their love-making they invited audiences to "become" them, to inhabit and expand their digital bodyscapes, to "hijack" their love story and turn it into a globally distributed community in "telematic embrace."[79] Throughout this performance, E8Z! and their audiences never met in physical proximity, never encountered each other through corporeal bodies. I believe that their mutual physical absence was a sine-qua-non for the performance. It was also the basis on which *Wirefire* could be built as an (unfinishable) performance piece, and a constantly in-flux, distributed, networked community "in love."

Closing this essay, I would like to refer back to Phelan's statement about the body being metonymic of presence and self in performance[80], and look at it from a networked, posthuman "pov." This brings forth a number of questions which I have attempted to articulate and explore – but which I have certainly not answered – such as: What constitutes "the body," and what constitutes "presence" in networked performance/performative encounters? How do we conceive of "self" and "other" in networked encounters – does this distinction still apply? Are binary strategies and dialectics such as presence-absence, materiality-immateriality and originality-artificiality pertinent within this context? Is the notion of the "original," auratic human body still relevant? Does the transformation of the "original," corporeal body into a hybrid, posthuman one entail absence (as opposed to presence) – or even a complete negation of the notion of the "self" as distinguished from the "other"? How can a contested self manifest its "presence"? Finally, has presence become indistinguishable from pattern and randomness?

I cannot offer answers to these questions, but there is one last thing I can offer: my fading memories of a few *Wirefire* encounters in which, as I saw it, performers and audiences had indeed

79. I refer to Roy Ascott's article "Is There Love in the Telematic Embrace?" *Art Journal*, 49,3 (1990): 241.
80. See this essay: "A *Wirefire* Encounter or the Presence-Absence Paradox."

ceased to be identifiable as self or other, present or absent. What I experienced were shared "bodyscapes" – that is, informational data and expanded materialities – that were constituted by the engagement between self and other, and manifested through fluctuating pattern(s) and randomness. In these encounters, as Stelarc has put it: "here and there, you and me, (were) meaningless distinctions..."[81]

81. http://www.stelarc.va.com.au/ (retrieved March 2006).

Blonde Birth Machines: Medical Simulation, Techno-corporeality and Posthuman Feminism

Jenny Sundén

New technologies of reproduction are rapidly transforming cultural understandings of kinship, family, body, sexuality and origin. Because of such changes, reproductive technologies have become a dynamic site of research across disciplinary boundaries. In the humanities and the social sciences, discussions of reproductive technologies interrogate the historical, political and cultural signification of technologically transformed parental definitions and practices, of new beginnings (and of endings), of the reconfiguration of the very limits of humanness and of "life itself." Techno-corporeality, the ways in which technologies are becoming (or perhaps, were always) entwined with our bodies, experiences and existences, comes in many shapes and guises in the domain of reproduction. Whereas considerable research has been devoted to technologies such as artificial insemination, *in vitro* fertilization (IVF) and "artificial womb" technologies,[1] considerably less attention has been paid to the design and use of birth machines in the field of medical simulation.[2]

1. See, for example, Adele Clarke, "Modernity, Postmodernity, and Reproductive Processes, ca. 1890-1990, or 'Mommy, Where Do Cyborgs Come From Anyway?'" in *The Cyborg Handbook*, ed. Chris Hables Gray, with the assistance of Heidi Figueroa-Sarriera and Steven Mentor (New York: Routledge, 1995); Robbie Davis-Floyd and Joseph Dumit, *Cyborg Babies: From Techno-Sex to Techno-Tots* (New York: Routledge, 1998); Sarah Franklin, *Embodied Progress: A Cultural Account of Assisted Conception* (London: Routledge, 1997); Sarah Franklin and Helene Ragoné ed., *Reproducing Reproduction: Kinship, Power, and Technological Innovation* (Philadelphia: University of Pennsylvania Press, 1998); Donna Haraway, "The Virtual Speculum in the New World Order," in *Revisioning Women, Health, and Healing: Feminist, Cultural, and Technoscience Perspectives*, ed. Adele Clarke and Virginia Olesen (New York and London: Routledge, 1999); Valerie Hartouni, *Cultural Conceptions: On Reproductive Technologies and the Remaking of Life* (Minneapolis: University of Minnesota Press, 1997).
2. Permission to re-publish this chapter has been granted by Ashgate. Originally, it appeared as: Jenny Sundén, "Blonde Birth Machines: Medical Simulation, Techno-corporeality and Posthuman Feminism," in *Technology and Medical Practice: Blood, Guts and Machines*, ed. Ericka Johnson and Boel Berner (Farnham: Ashgate, 2010), 97-118. This text is part of the research project *Technologies of Life: On Art, Science and*

Meet S575 Noelle™ – a wireless maternal and neonatal computer interactive simulation system from Miami-based Gaumard Scientific – one of the latest incarnations of cutting-edge birthing machinery.[3] S575 Noelle™ (pronounced as *Noël*, referring to its release near Christmas) is delivered with: one fully computer interactive and automated delivery system (the manikin body assembly); placenta with umbilical cord; episiotomy trainer (set of three); urinary bladder; a lid covering the abdomen (and a special lid for C-sections); a lifelike birth canal, postpartum haemorrhage; one full-term neonate; interface modules for ECG, pulse, ultrasound, defibrillation, cardioversion, and pacing of Noelle and neonate; two 17-inch touch-screen monitors; one computer for the instructor to control the monitors and communicate wirelessly with the system; one student laptop to control fetal delivery, neonate condition and the monitors; instruction manuals and teaching tips.

Bringing out the age-old Frankensteinian fear of technology-out-of-control in its merging of technology with the female body, Noelle is designed to give a number of birth scenarios.[4] It can be

New Media, financed by The Swedish Research Council (Vetenskapsrådet).

3. Gaumard Scientific traces its history to 1946. The founder was a World War II army physician, George Blaine, who worked on reconstructive surgery and childbirth technologies. One of the first projects he introduced in 1949 was a childbirth simulator: "The body of the simulator was approximated by translucent thermoplastic, the abdomen by padded cotton, the vulva was approximated by two hemispherical pieces of latex, and the baby was a rag doll" (Interview with Vice President of Gaumard, John Eggert, November 22, 2006). Gaumard has developed and delivered Noelle models since 1999, the "575" being the latest incarnation from 2008.
4. In Mary Shelley's *Frankenstein: Or, the Modern Prometheus* (London: Penguin Books, 1818/1994), the monster's tragedy is his solitude, his exclusion from human relations in general, and from an intimate connection with a female counterpart in particular. He begs Victor to create a female monster, but halfway through her construction, Victor decides to destroy her. While the monster vengefully watches him through the window, Victor violently terminates his second act of (pro)creation, since this female monster "might become ten thousand times more malignant than her mate," she may even "become a thinking and reasoning animal." The monster had promised to hide from human society, but she had not. There was no guarantee that she would voluntarily adapt to the subordination of femininity. Moreover, the monster may hate her deformed body more than he hates his own, or, perhaps even worse, she "might turn with disgust from [the monster] to the superior beauty of man." (Shelley, *Frankenstein*, 160). This fear is a central theme in James Whale's horror film *The Bride*

in labour for hours or give birth unexpectedly fast, and be programmed to display a range of different birth complications. In his article, "Robot birth simulator gaining in popularity," Associated Press business writer Paul Elias mentions how this "full-sized, blond, pale mannequin is in demand because medicine is rapidly abandoning centuries-old training methods that use patients as guinea pigs, turning instead to high-tech simulations."[5] Noelle is a first-rate high-tech simulation, eventually delivering a plastic doll – a baby robot – that can change colors "from a healthy pink glow to the deadly blue of oxygen deficiency ... wired to flash vital signs when hooked up to monitors."[6] Both "mother" and "baby" produce realistic pulse rates and are able to breathe. The latest edition to the Noelle series is simulated blood in a uterine assembly, attached under the lid of the abdomen. When a port is opened, this produces a blood-like fluid from the inside of the uterus, the walls of the birth canal or both. An instructor can reduce bleeding as students perform fundal massage. Noelle varies in price, ranging from $3,200 for the most basic model to a $38,000 fully interactive, wireless version. The motto for the wireless Noelle is to simulate "care in motion."

Departing from a posthuman feminist understanding of technobodies, this chapter critically explores simulations of birth, the lifelike, and techno-corporeality through a close encounter with a (primarily) blonde, (primarily) white birthing machine. The text draws on a range of sources, notably technical manuals, instruction videos, as well as an interview with the simulator's "found-

of Frankenstein (1935). Mary Jacobus (in "Is There a Woman in this Text?" *New Literary History, A Journal of Theory and Interpretation* XIV, 1 (1982): 117-141) points out that the fantasy of the female monster who longs for men instead of monsters is a terrifying threat to male (hetero)sexuality. This threat is not only tied to the she-monster's hideous deformity that excludes his image of desire, but to the frightening autonomy of her refusal to reproduce in the image in which she was made. For a development of this argument, see Jenny Sundén, "What if Frankenstein('s Monster) was a Girl? Reproduction and Subjectivity in the Digital Age," in *Bits of Life: Feminism at the Intersections of Media, Bioscience, and Technology*, ed. Anneke Smelik and Nina Lykke (Seattle, University of Washington Press, 2008).

5. Paul Elias, "Robot birth simulator gaining in popularity," *Associated Press*, 2006, accessed: June 18, 2010, http://www.usatoday.com/news/health/2006-04-15-robot_x.htm.
6. Elias.

ing father," but it also take into account technological imaginaries of simulator bodies in terms of how the simulator is imagined in popular science contexts. The focus of this exploration is on the design of simulators, and not on actual use in clinical settings (which would be the subject for a different investigation). Then again, thinking about design for medical practices is thinking about use in the sense that design processes and practices always inscribe and anticipate use and users in certain ways (and not others). An investigation of techno-corporeal imagery and becomings in medical simulation, their possibilities and limitations, the text critically addresses the politics of simulation.

The argument is performed in three parts. It sets out with an account of the "birth" of birth simulators themselves – the act of creation in a domain of technological procreation – investigating the status of the "real" as well as of "realism" in the simulation world. Second, it moves on to explore some historical parallels or predecessors to medical simulators in the shape of anatomical wax models. Finally, in response to recent feminist discussions of difference in terms of "intersectionality" (how various power relationships co-construct one another in multiple ways), the discussion will reveal how this particular reproductive machinery is entwined with issues not merely of sexual difference and sexuality, but also of race and national belonging.[7] The chapter, starting with the birth of simulators, ends, conversely, with a brief contemplation of simulation and death.

Design Beyond Realism

In the 1990s, people at Gaumard realized that they needed to take their work on motherhood and birthing experience in a different direction. While they had previously worked with models of the female body in parts, such as obstetric phantoms, they now envisioned that models of birthing needed to be divided into three elements, namely, the mother, the foetus and the neonate. On 22

7. For a discussion of the possible affinities and affordances linking the current intersectionality discussion to the field of cyber/cyborg-feminism, see Jenny Sundén, "On Cyberfeminist Intersectionality," in *Cyberfeminism in Northern Lights: Digital Media and Gender in a Nordic Context*, ed. Malin Sveningsson Elm and Jenny Sundén (Cambridge Scholars Press, 2007).

November 2006, I conducted an online interview with John Eggert, the Vice President of Gaumard Scientific and an engineer with a long history in medical simulation.[8]

> It became fairly obvious that we needed a full-sized mum that would articulate properly. We needed a fetus that could be birthed either vertex or breech or anything else. And we needed to have a neonate that would respond appropriately ... And then, it is pretty apparent that you'd need a mechanical solution, and you'd need an electrical hardware solution, and you'd need a software solution to conduct the orchestra.[9]

Such a complex cybernetic machinery of mechanical, motor-driven hardware bodies, interlinked with, displayed, monitored and operated by screen-based scenarios, is certainly a challenging design project. In medical debates about the use of simulators in medical education, there is a clear emphasis on the need for cost-effective, low-risk, educational strategies to meet all the more complex technological development, paired with a heightened awareness of patient care and safety.[10] Traditionally, the psychomotor skills and clinical reasoning skills required to be a clinician were acquired via an apprentice-style "see one, do one, teach one" model. Today, medical simulators are introduced as mediators and problem-solvers, taking the place of humans as experimental subjects.

8. Eggert situates himself as a creative father of sorts of Noelle, even if the simulator is an ongoing, collaboratively designed configuration. It was Eggert who responded to my initial email contact with Gaumard, investigating the possibilities of interviewing people from the Noelle design team: "I am the creator. How can I help?" Eggert also performs alone together with the simulators in the instruction videos on Gaumard's homepage, accessed: June 18, 2010, http://www.gaumard.com/index.html (under "Videos").
9. Interview: John Eggert, November 22, 2006.
10. See, for example, Shad Deering et al., "Simulation Training and Resident Performance of Singleton Vaginal Breech Delivery," *Obstetrics & Gynecology* (2006): 107; Lambert W. T Schuwirth and Cees P. M. van der Vleuten, "The Use of Clinical Simulations in Assessment," *Medical Education* (2003): 37; Melanie C. Wright et al., "The Use of High-Fidelity Human Patient Simulation as an Evaluative Tool in the Development of Clinical Research Protocols and Procedures," *Contemporary Clinical Trials* (2005): 26.

Discussions of the usefulness of simulators enter in intriguing, and not always consistent, ways into the slippery terrain of "realism," "the lifelike" and "the real." Realism, for example, as a literary mode is no less constructed than say fantasy or concrete poetry, but is harder to reveal as such. Literary realism, seemingly operating as a transparent mode, "is characterized by an inability or a refusal to know itself as writing; illusionism, a kind of willful self-blindness destined to induce in its reader a reciprocal self-blindness."[11] Realism, traditionally associated with literature from the Renaissance and on, is characterized by its focus on individual experience and bounded by its mission to represent the "real" conditions of social existence:

> Realism involves a fidelity both to the physical, sensually perceived details of the external world, and to the values of the dominant ideology. ... Realism's desire to "get the details right" is an ideological practice, for the believability of its fidelity to "the real" is transferred to the ideology it embodies. The conventions of realism have developed in order to disguise the constructedness of the "reality" it offers, and therefore of the arbitrariness of the ideology that is mapped onto it.[12]

The desire in realism to "get the details right" has an obvious resonance in the simulation world. On one hand, simulation is said not to be intended to replace learning in clinical settings, but rather to provide a controlled and safe environment for practising. "As such, simulation imitates, but does not duplicate reality – it is a controlled "real-world-like" medical setting" in which the simulator body mimics the processes of a living body: a beating heart (or, actually, several), internal movements, and rotations, and so on.[13] On the

11. Penny Boumelha, "Realism and the Ends of Feminism," in *Grafts: Feminist Cultural Criticism*, ed. Susan Sheridan (London and New York: Verso, 1988), 81.
12. John Fiske, *Television Culture* (London: Methuen, 1987), 36.
13. V.A. Ypinazar and S.A. Margolis, "Clinical Simulators: Applications and Implications for Rural Medical Education," *Rural and Remote Health* 6 (2006): 3, accessed June 18, 2010, http://www.rrh.org.au/articles/subviewnew.asp?ArticleID=527.

other hand, there is a desire to create scenarios that masquerade as "reality," that take the place of "the real" in ways that make the participants forget about its constructedness. This understanding approaches Baudrillard's notion of simulation as that which does not pretend, but produces itself as that which is simulated, hence rendering unclear the difference between true and false, real and imaginary.[14] Or, in Eggert's words (words that recur like a mantra in the simulation community): "Isn't the classic phrase that you want to suspend disbelief?"[15] In this sense, simulators are thought of as mimetic devices, as stand-ins for "the real," as critically close approximations. If being critically close has bearing on discussions in medical simulation, a relevant question is then to what, more exactly, such critical closeness refers. In other words, what is positioned as and understood to be "the real" in medical simulation?

As feminist science and technology scholar Ericka Johnson points out, there are typically double concerns in discussions in the simulation world as to what constitutes a successful, or "valid," simulator: Does the simulator mimic a real body realistically? Does it teach medical procedures accurately?[16] The method commonly used to evaluate simulators is double blind clinical trials. Rather than relying on measurements of whether simulator bodies modelled in "lifelike" ways have significant impact, these tests speak primarily of how well the simulators mediate the knowledge and experience of medical practices.[17] Johnson argues that while medical simulators do provide models of the patient body, the body being modelled in the simulator – rather than constituting an ana-

14. Jean Baudrillard, *Simulacra and Simulation*, trans. Sheila Faria Glaser (Ann Arbor: University of Michigan Press, 1994).
15. Interview: John Eggert, November 22, 2006.
16. Ericka Johnson, "Simulating Medical Patients and Practices: Bodies and the Construction of Valid Medical Simulators," *Body & Society* 14, 3 (2008): 105-128.
17. See, for example, Deering et al., "Improving Resident Competency in Management of Shoulder Dystocia With Simulation Traning," *The American College of Obstetrics and Gynecology* 103, 6 (2004); Halamek et al., "Time for a New Paradigm in Pedriatric Education: Teaching Neonatal Resuscitation in a Simulated Delivery Room Environment," *Pedtriatrics* 106 (2000): e45, accessed June 18, 2010, http://www.aap.org/nrp/simulation/newparadigm.pdf.; Melanie C Wright, Jeffrey M. Taekman and Mica R. Endsley, "Objective Measures of Situation Awareness in a Simulated Medical Environment," *Quality and Safety in Health Care*, 13 (2004).

tomically pre-existing given – is actually the *experienced* body. Her main point is that medical simulators, more than anything else, articulate medical practices, which begs critical questions concerning whose practice, and experience, is being simulated.

When asked about what kind of competence, knowledge and experience was constitutive in designing Noelle, apart from the most obvious ones in the mechanical engineering and programming fields, Eggert spoke about the accumulative, dynamic nature of simulator design and the importance of taking into account the company's history of having designed, developed and sold childbirth simulators for fifty years. Implicit in this history is, for example, extensive customer input from people around the world, which becomes a knowledge base for members of the company. Then again, he also delivered an interesting answer that has everything to do with family bonds and reproductive capacities:

> JE: We have on our staff two physicians who work with me on a regular basis. One of which is my son. He is an internist, pulmonologist and critical care specialist.
> JS: Did you bring in people from the other side of the spectrum, so to speak, like women who have experience of giving birth?
> JE: Those that we brought in were and are still in the company, and would include people like my wife.

It is imperative to point out that the Noelle project provides multiple models of practised bodies and bodies as practices, so to speak. The Noelle body has never been one, but is the (provisional) result of multiple bodies and body-technologies, a cyborgian assemblage of an ongoing process of modelling and remodelling, a continuous transformation. In Eggert's words, "the Noelle project, which we started years ago, in fact will never be completed."[18] Designing for medical practices is itself a practice, as well as a kind of use, in the sense that design practices inscribe and anticipate use. Given the open-endedness of design for medical practices, and how it must anticipate the use of the simulator, this process is certainly not with-

18. Interview: John Eggert, November 22, 2006.

out its limitations. Rather, what the simulator body can do, is regulated, restricted and controlled in multiple ways. As will become clear, such restrictions have everything to do with how simulator design practices incorporate the idea(l)s of technological progress and a kind of embodiment that is peculiarly disembodied.

As important as it is to try and trace the bodies, models and technologies that, in their way, shape the simulation, it is equally important to understand that this is never a one-way process – as seems to be the underlying assumption in the rhetoric of realism and simulations of "the real." Alongside Lisa Cartwright, I find it crucial to ask "whether conventions of realism are in fact the most useful means of "learning more" about, or "better representing" the body" in medicine.[19] Moving beyond realism to better understand medical simulation, however, is not merely a matter of revealing how realism masquerades as the universal, and hence conceals the cultural specificity and situatedness of technologies. This is because such discussion keeps insisting on a division between the real and the simulated, between biological bodies and those of machines, even in *critiquing* the principles of realism. Even if the universalism of realism in simulation is questioned by asking whose reality, more precisely, is being simulated, the notion of a pre-existing reality possible to re-present and model (with the simulator) remains. Such an understanding obscures the fact that there is no clear-cut causality between human and nonhuman bodies, no obvious referential link.

Techno-Corporeality and Posthuman Feminism

If (human) bodies are thought of as pre-existing simulation, as originals of sorts against which the simulators are measured as more or less successful incorporations, every alternative understanding of the many connections between human and nonhuman bodies, between nature and culture, is effectively eliminated.

Feminist science and technology scholars have consistently decoupled nature from the natural in showing that "nature,"

19. Lisa Cartwright, "A Cultural Anatomy of the Visible Human Project," in *The Visible Woman: Imaging Technologies, Gender, and Science*, ed. Paula A. Treichler, Lisa Cartwright, and Constance Penley (New York: New York University Press, 1998).

much like culture, is something that is formed and re-formed. This decoupling has also served to disengage the die-hard coupling of woman and nature, or woman-as-nature. Interestingly, women and nature appear to be re-coupled in much feminist work on reproductive technologies. Reproductive technologies – whether understood as artificial insemination, *in vitro* fertilization, "artificial womb" technologies or ultrasound scanning – are often articulated independently of/as separate from women, their experiences and (supposedly) natural bodies. As Irina Aristarkhova has pointed out, "very few feminist researchers to date have articulated other than the most negative attitudes toward "reproductive machines," mostly on the grounds that women are already being exploited as "wombs" and "machines of reproduction."""[20]

It is not surprising that feminist critics of reproductive technologies are attempting to reclaim women's bodies as a strategy against what is generally referred to as the medicalization of women's bodies, as well as against tendencies to exclude the maternal from the procreational equation. In these discussions, the body of the woman/mother is violently invaded, inscribed, disciplined or controlled by (masculine) medical science and technologies. Alternately, their bodies and their reproductive capacities are sidestepped altogether and ultimately replaced by machines.

In feminist criticism of technoculture more generally, there is a similar line of thought that emphasizes how the (sexed) nature/culture split prevails even in intense fusions of human and machine. Machine bodies in Hollywood science fiction, as pictured in films such as *Robocop* and *The Terminator*, typically illustrate an exaggerated, metallic masculinity that highlights traditional couplings between male bodies and machinery.[21] Although a synthesis

20. Irina Aristarkhova, "Ectogenesis and Mother as Machine," in *Body & Society* 11, 3 (2005): 52. For discussions that clearly complicate the techno/body division, see, for example, Haraway (1991), Haraway (1999). See also Charis Thompson's more optimistic take on technologies of reproduction as new possibilities for women, in *Making Parents: The Ontological Choreography of Reproductive Technologies* (Cambridge MA: MIT Press, 2005).
21. On the close connection between masculinity and science and technology, see, for example, Cynthia Cockburn, *Brothers: Male Dominance and Technological Change* (London: Pluto, 1983); Cynthia Cockburn, *Machinery of Dominance: Women, Men, and Technical Know-How* (London: Pluto, 1985); Ruth Oldenziel, *Making Technology Masculine: Men,*

of human and machine that renders unstable the very boundaries of the human subject, the maleness of these cyborg bodies is not only left intact, but hyped-up.[22] Even if such readings of machine bodies rightfully draw attention to how the sexed body does not necessarily vanish out of sight in corporeal linkages with technologies, they seem to sidestep, entirely, the fact that cyborg bodies are haunted by contradictions. If (reproductive) technologies are, too hastily, seen as a male domination/denial of repressed female bodies, this builds on the problematic notion that there would exist such a thing as a (feminine) sphere untouched by technology. Jill Marsden points out how such commentaries imply a "separation or distinction between technological "agency" and the "matter" of its operations"; in this sense, it then becomes possible to speak of "natural" nature, either irresponsibly wished away (as in the Frankensteinian death of the mother) or ideologically determined (as a resource for masculine exploitation and control).[23] Consequently, these arguments effectively eliminate every productive pairing of *women* and machines. They also eliminate an understanding of the materiality of the body, not as passive or culturally determined, but as an active force of its own becoming.

Elsewhere, I have argued for how Australian "corporeal feminism" as formulated by feminist philosopher Elizabeth Grosz can be productively put to use in cyborg-feminist projects.[24] In relation to Grosz, for whom the body is understood as something partial, incomplete, relational and always in the process of be-

Women and Modern Machines in America, 1870-1945 (Amsterdam: Amsterdam University Press, 1999); Judy Wajcman, "Technology as Masculine Culture," in *The Polity Reader in Gender Studies*, ed. A. Giddens, D. Held and D. Hillman (Cambridge: Polity Press, 1994); Judy Wajcman, "Feminist Theories of Technology," in *Handbook of Science and Technology Studies*, ed. Sheila Jasanoff (Thousand Oaks, CA: Sage, 1995).

22. Samantha Holland, "Descartes Goes to Hollywood: Mind, Body and Gender in Contemporary Cyborg Cinema," in *Cyberspace, Cyberbodies, Cyberpunk: Cultures of Technological Embodiment*, ed. Mike Featherstone and Roger Burrows (London: Sage, 1995); Claudia Springer, "The Pleasure of the Interface," *Screen* Autumn (1991); Claudia Springer, *Electronic Eros: Bodies and Desire in the Postindustrial Age* (Austin, TX: University of Texas Press, 1996).
23. Jill Marsden, "Virtual Sexes and Feminist Futures: The Philosophy of 'Cyberfeminism,'" *Radical Philosophy* 78 (1996): 8.
24. See Jenny Sundén, *Material Virtualities: Approaching Online Textual Embodiment* (New York: Peter Lang Publishing, 2003).

coming otherwise, but nonetheless as always sexually specific, the discussion of women and/as reproductive machines could take a different direction. In *Sexy Bodies: The Strange Carnalities of Feminism*, Elizabeth Grosz and Elspeth Probyn argue for a mode of understanding (the sexing/queering of) bodies along a Deleuzian notion of becoming; of bodies conceived as activities, processes, movements, lines and motions of making.[25] This is not the time or place for a thorough exposition of the possible affinities and implications of corporeal feminism for feminist science and technology studies, but merely an opportunity to point out a possible direction for such an endeavour. For Grosz, the body is neither a pre-representational biological presence, nor a purely discursive inscription, but an "open materiality" and a site for political, social and cultural struggle:

> It is an open materiality, a set of (possibly infinite) tendencies and potentialities which may be developed, yet whose development will necessarily hinder or induce other developments and other trajectories. These are not individually or consciously chosen, nor are they amenable to will or intentionality: they are more like bodily styles, habits, practices, whose logic entails that one preference, one modality excludes or makes difficult other possibilities.[26]

In thinking bodies in terms of becoming, there is the possibility of distinguishing between different modes of becoming, between different ways in which bodies become meaningful. Design for medical practices could, in similar terms, be thought as a set of potentialities which may be developed, and yet whose development will hinder other ways in which simulations may take shape. In an attempt to challenge the boundaries of constructivism, an exploration of the technological and cultural shaping of bodies is not enough. It is important also to examine how sexually specific yet

25. Elizabeth Grosz and Elspeth Probyn ed., *Sexy Bodies: The Strange Carnalities of Feminism* (London and New York: Routledge, 1995).
26. Elizabeth Grosz, *Volatile Bodies: Toward a Corporeal Feminism* (Bloomington and Indianapolis: Indiana University Press, 1994), 191.

hybrid bodies come to matter, and will matter differently, partly due to their very materiality. If Grosz's early work was an effort to think the body as that which marked discourse with its own force and difference, in her later work on Bergson and Darwin, Grosz expands this site of force not only beyond the body, but beyond the human altogether.[27] This is a relocation of the argument, from its previous investment in human bodies and/as culture, to a heightened interest in intersections of nature and culture, life science and cultural theory. This reorientation, then, also potentially expands the discussion of corporeality beyond the bodies of humans, to also include non-human bodies and matter. In a similar way, feminist physicist Karen Barad argues that feminist philosophy of the body (in her case primarily Butler's notion of performativity) can be productively de-anthropomorphized and put to use for science studies.[28]

This shift of emphasis, from an interest merely in culture, to a focus on intersections of nature and culture, could imply a willingness to let a feminist thinking of bodily becomings speak to relationships between human and nonhuman bodies and to how they are implicated in one another. What I am getting at is a *posthuman feminist* understanding of techno-corporeality. Along with Barad, in her definition of a posthumanist perspective, I argue for feminist interventions and interrogations, which "[call] into question the givenness of the differential categories of "human" and "nonhuman," examining the practices through which these differential boundaries are stabilized and destabilized."[29] In thinking with and through reproductive machines, it is important to recognize how these technologies, rather than being something external to and possibly invasive of the body, are actually connected in multiple ways with what we have come to think of as our bodies. In her interrogation of the temporality of the technobody, Gail Weiss speaks of the interface between bodies and machines, of the "significance of an *intercorporeality* that defies any attempt

27. See, for example, Elizabeth Grosz, *The Nick of Time: Politics, Evolution, and the Untimely* (Durham and London: Duke University Press, 2004).
28. Karen Barad, "Posthumanist Performativity: Toward an Understanding of How Matter Comes to Matter," *Signs* 28, 3 (2003).
29. Barad, "Posthumanist Performativity," 808.

to affirm the autonomy of the body apart from other bodies or from the disciplinary, technological practices that are continually altering and redefining them."[30] She traces the relationality between bodies and between bodies and machines in temporal terms and envisions the temporality – and, indeed, the embodiment – of technobodies as nothing other than our own:

> The durée of the technobody – whether this body be that of a newly cloned sheep; "a test-tube" baby; a woman in labour hooked up to technological devices that record fetal movement, fetal heartbeat, maternal blood pressure, and contractions; a boy with a liver transplant; a woman with a prosthetic leg – arises out of a violent effort and requires a violent effort in order to see the interconnections that link this durée with our own. Indeed, what is most violent, I think, is the recognition that technology is part and parcel of our own durée.[31]

The question is not whether or how women can be saved from (reproductive) technologies, but one of working with and through our couplings with machines responsibly. To acknowledge the relationality between human and nonhuman bodies is not to give in to a certain masculine script of technological development and progress, but to envision the only site possible for intervention, infiltration and possible redesign.

Woman as Birth Machine

Woman as enunciated by the patriarchy as a birth machine, predestined for the reproduction of the labour force in capitalist so-

30. Gail Weiss, "The Durée of the Technobody," *Becomings: Explorations in Time, Memory, and Futures*, ed. Elizabeth Grosz (Ithaca: Cornell University Press, 1999), 163. See also Catherine Waldby's use of Weiss's term *intercorporeality* to address notions of embodiment and tissue transfer. She argues: "the idea of intercorporeality contributes to a denaturalization of the relations between the limits of the body and the limits of the 'I' understood as a discrete entity. In this regard, it may help to conceptualize that most literal kind of boundary confusion involved in tissue transfer." In Catherine Waldby, "Biomedicine, Tissue Transfer, and Intercorporeality," *Feminist Theory*, 3, 3 (2002): 241.
31. Weiss, "The Durée of the Technobody," 170.

ciety (where men are the real source of capital), is far from a new figuration. In her writing on scientific disembodiment, pregnancy and the unborn, medical historian Barbara Duden argues that human foetuses today, as well as (pregnant) women, rather than being conceptualized as natural facts are engineered constructs of modern society.[32] To this, I would add that the notion of being engineered also entails the possibility of re-engineering, of being formed differently, an argument rarely present in the discussion of the "technologization" of women's bodies.

Reproductive machines have circulated in Western cultures in many shapes and guises. Medical simulators and birthing machinery, such as Noelle, come out of a long tradition of, for example, obstetric phantoms and anatomical wax models. Wax models – or waxes – were introduced as didactic tools toward the end of the seventeenth century, and remained in use well into the eighteenth century. Apart from their educational purposes, they also made for thrilling public displays on the boundary of art and medical science. In Ludmilla Jordanova's account of wax models, she puts forth how these "Venuses" (as they were called), shipped out from northern Italy "on silk or velvet cushions, in passive, yet sexually inviting poses," were intriguing modulations of art, medical science, reproduction and eroticism.[33] Even if not all waxes were female, those that were male were most often either upright, fleshless muscle men, or merely parts of men, such as male torsos.

According to Jordanova, the differences between male and female waxes are even more striking in view of how the female bodies of wax are detailed, adorned and supplemented. With minute attention to detail, the bodies of the female wax models resemble Roman statues in how they depict the female shape. Instead of the smooth, stone-like impression of cold marble, the use of wax

32. Barbara Duden, *Disembodying Women: Perspectives on Pregnancy and the Unborn*, translation Lee Hoinacki (Cambridge, MA: Harvard University Press, 1993).
33. Ludmilla Jordanova, *Sexual Visions: Images of Gender in Science and Medicine Between the Eighteenth and Twentieth Centuries* (New York: Harvester Wheatsheaf, 1989), 44. For discussions of female wax models as erotically inviting, see also Elisabeth Bronfen, *Over Her Dead Body: Death, Femininity, and the Aesthetic* (New York: Routledge, 1992); Elaine Showalter, *Sexual Anarchy: Gender and Culture at the fin de Siècle* (London: Bloomsbury, 1991).

painted with oil colours, so it can be mistaken for human skin, makes for an uncanny realness.[34] Human hair frames their faces and covers their sexes; they have body parts that can be removed, revealing the texture underneath the skin, and are sometimes supplemented with small foetuses. Many wax Venuses strike poses from famous art works (as their name suggests), which not only links art to science, but points, according to Jordanova, to realism as a guiding principle in multiple fields:

> Eyebrows, lashes, head and pubic hair were added painstakingly and serve no other function than to make the body as lifelike as possible. They add nothing to the anatomical detail of the model; they even form a bizarre and striking contrast to the exposed internal organs and muscles that look like chunks of meat. Here we have more than realism; a verisimilitude so relentless that is becomes hyper-realism.[35]

Attention to detail is everything in simulation, which brings Noelle in line with previous displays of the reproductive female body in medicine, yet in a different way. Jordanova distinguishes between lifelikeness and anatomical detail and argues that the adorned body of the waxes had nothing to do with how these bodies model anatomy, with how they function as didactic tools. In Noelle, there is perhaps a similar contrast between, for example, the human-like hair on the head and the internal motor rotations that gradually push out the baby. But even if perfectly washable,

34. Departing from Sigmund Freud's 1919 essay "The Uncanny", the anthology *The Uncanny: Experiments in Cyborg Culture*, edited by Bruce Grenville (Vancouver: Vancouver Art Gallery, Arsenal Pulp Press, 2001) is an investigation of cyborg imagery and cyborg bodies. In the introductory chapter, Grenville uses Freud's understanding of the uncanny as something not new or strange, but frighteningly familiar, to understand today's cyborgs: "I argue that the cyborg is uncanny not because it is unfamiliar or alien, but rather because it is all too familiar. It is the body doubled – doubled by the machine that is so common, so familiar, so ubiquitous, and so essential that it threatens to consume us, to destroy our links to nature and history, and quite literally, especially in times of war, to destroy the body itself and to replace it with its uncanny double." (Grenville, *The Uncanny*, 21).
35. Jordanova, *Sexual Visions*, 47.

combable simulator hair adds nothing as such to the process of delivery, it feeds into and foregrounds affective responses on the part of the human user. In performing birth and birthing "as if" the nonhuman body of the simulator is, in fact, human, attentiveness to detail in the simulator body is not set apart from medical practices, but becomes an intrinsic part of them. Elias describes a training session in which, in a state of emergency, the Noelle was wheeled off to an operating room "where she gave regular birth to twins after a frenzied 20-minute operation."[36] One of the nurses expressed concerns as to whether or not the simulator body was appropriately covered ("We wheeled her through the hallway with her gown open"), effectively destabilizing the distinction between simulator and simulated, nonhuman and human bodies.[37]

Then again, ways of depicting female bodies for "medical purposes" do not always relate to concern with dignified cover-ups. On the contrary, bare naked waxes were put on display as still lives for an audience to learn from, and to be thrilled and perhaps aroused by. Jordanova makes the case that the female wax models invited sexual thoughts in the form of "mentally unclothing the woman. Of course, these models were already naked, but they gave an added, anatomical dimension to the erotic charge of unclothing by containing removable layers that permit even deeper looking into the chest and abdomen."[38] Noelle is "layered" in a similar manner as the waxes, allowing the medical team to remove parts and, in a sense, look into the interiority of the body and its reproductive machinery. Even though there are occasional news photos of Noelle clad in a hospital gown, most of the time the Noelle simulator is on display completely naked.

The nakedness of Noelle's body is even more striking in comparison with some of its male counterparts, such as the emergency (or EMS) simulator SimMan from Norwegian Laerdal. SimMan is routinely portrayed on the company website wearing pants and a shirt, sometimes with one sleeve rolled up over the elbow to facilitate the insertion of IV needles.[39] If not fully clad,

36. Elias.
37. Ibid.
38. Jordanova, *Sexual Visions*, 55.
39. See, for example, the first page of the 2007 Product Catalogue from

in news shots from training sessions around the world, SimMan may appear bare from the waist up, while visual evidence of the whole body laid bare is rare. Disrobing Noelle, then, becomes not a process of removing garments, but rather a matter of removing and/or replacing body parts, such as the "lid" covering the abdomen, the cervix (which often breaks), and so on. As Eggert points out in a discussion of the design and inter-changeability of body parts, "the cervix is an enormously difficult design exercise," since the material used needs to be enormously elastic, feel soft when touched, yet have longevity.[40] In comparison with, for example, an arm, cervixes are not long-lived: "the cervixes are replaceable, they snap in and out ... Whereas arms literally want to be in the field for years without having to be replaced or without having significant maintenance done on them."[41]

In contrasting eyebrows and pubic hair with "the exposed internal organs and muscles that look like chunks of meat," Jordanova traces a representational shift from "realism" to "hyper-realism." Medical simulators like Noelle similarly seem to enter, not realism, but rather a (materialized) Baudrillardian hyper-realism in ways that, while challenging the distinction between human and nonhuman bodies, produces a hyper-real body that is curiously *dis*embodied.[42] This techno-corporeal formation relies on a selective bodily awareness that allows the birthing body to have hair that can be washed and combed, breasts perfectly shaped, and so on, but that discriminates quite clearly against bodily fluids, odours and sounds. Even if Gaumard provided bodily fluids thirty years ago, the typical customer response would be that the less messiness, the better, to avoid having to bother with any clean-up. In terms of audio, screaming is an add-on choice of sorts, performed by the instructor. It is a scream wired into the body from the outside, but not something clearly linked with certain stages or incidents in the delivery phase. In line with customer requests, and in line with the idealization of female bodies in technoculture,

Laerdal, accessed: June 18, 2010, http://www.laerdal.no/document.asp?docID=14668676.

40. Interview: John Eggert, November 22, 2006.
41. Ibid.
42. Baudrillard, *Simulacra and Simulation*.

the Noelle comes together as a rather dry, quiet, odourless configuration.[43] Human wet-ware has primarily entered the simulator indirectly, such as in the instruction manual reminder to lubricate the birth canal before each delivery. Interestingly, one of the latest editions to the Noelle simulator series is simulated blood postpartum, which points at a change of attitudes on the part of the customers. Another interesting development tendency is the change from a relatively quiet delivery (Noelle had previously no speech), to models with nearly a hundred pre-programmed phrases to be initiated by the instructor. Noelle's speech concerns everything from experiences of pain and discomfort, to comments on how the pregnant body has felt recently. But if SimMan can beg for his life ("Don't let me die"), and draw attention to the fact that he is not alone ("I have a family"), the Noelle scenarios rarely get more dramatic than "worst pain," "going to be sick" and "help me."

The Detachable Feminine And Simulations Of Whiteness

A virgin birth machine, the Noelle body is not only one, but at least two. In being delivered with a foetus as well as a neonate of indeterminate sex, Noelle is many. However, this possibly frightening multiplicity in the crossing of wires and reproductive sexuality seems to be taken care of by the male engineer with god-like master commands at his fingertips: "David Isaza, an engineer with Gaumard, sat in a corner with a laptop, sending wireless signals to Noelle. With a keystroke, he can inflict all sorts of complications, overriding any preprogrammed instructions."[44] Noelle, with its high-wired womb and always ready to dilate cervix and vulva, is in a sense the techno-incorporation of a new, automaton

43. The dryness of this cyborg body parallel in a sense the dream of techno-disembodiment, of leaving the body and all its limitations behind in cyberspace articulated by (primarily male) cyberenthusiasts in the early 90s. In their Cartesian separation of mind from body, cyberspace is completely freed from the messiness of the physical and miraculously becomes a space of the mind. The dream of getting rid of the all too human wet-ware in simulator birthing bodies is of course not the same, but still carries a similar tendency of not blending high-tech, computer interactive bodies with the messiness of human embodiment.
44. Elias.

virgin Mary, giving birth to the One endlessly. Interestingly, the newborn robot is peculiarly sexless, its apparently seamless crotch avoiding every straightforward linguistic determination. Is this a pointer in the direction of a high-tech, queer future open to ambiguity and trans-sexuality? Or, is it rather a way on the part of the mannequin designers of not having to pick sides?

The answers to these questions are perhaps not as clear-cut as they may at first appear, at least not in relation to SimMan, which puts an interesting spin on the simulation of reproductive heterosexuality. At first sight, SimMan simulates the reliable breadwinner and family father (through uttering canned phrases such as "I have a family"). In the next instant, however, SimMan may come across as a thought-provoking trans-sim. An article entitled "Dolls give students a dose of reality" reports on how interchangeable body parts of simulators put a twist on bodily specificity in simulation. Nursing senior Sharon Williams at the University of Pennsylvania recalls how one time she had to insert a catheter in one simulator and was met with a surprise: "I think I am doing it on a male because the doll had a male face," she says. "I pulled down the sheet, and there was a female part!"[45] Another report clarifies that "SimMan can become a woman, with the addition of a wig, breasts and reproductive organs."[46]

In a true transvestite, drag queen fashion, the "he" becomes a "she" with the *addition* of (female, one must assume) reproductive organs, breasts and, not to forget, a *wig*. This makes one wonder whether some sort of male reproductive organs have to come off first, before anything else can be added. Or, if this sexed transformation rather proceeds from a non-genital, non-reproductive default mode, to which "the feminine," in the form of reproductive machinery, is a possible add-on. At moments like these, it is quite striking how the production of the simulator body is curiously queer – even in the midst of (heterosexually infused) reproduction.

45. Melissa Leiman, "Dolls give students a dose of reality," *The Daily Pennsylvanian*, January 20, 2003, accessed June 18, 2010, http://www.temple.edu/ispr/examples/ex03_01_20b.html.
46. In *Rounds*, Summer 2003, Hartford Hospital's wellness magazine, accessed December 15, 2006, http://www.harthosp.org/rounds/PDF/2003.

Even though the Noelle model comes in "three ethnic colours," Gaumard offering what Eggert refers to as "a Northern European, a Southern European and an Afro," as a multicultural solution to the simulation of racial difference, the simulator is routinely portrayed and sold as white.[47] In her *White Women, Race Matters*, Ruth Frankenberg makes clear that whiteness is both a privileged location, a perspective from which to understand the self and others, and a set of cultural practices that are usually unmarked and unnamed: "Naming whiteness displaces it from the unmarked, unnamed status that is itself an effect of its dominance."[48] The naming of whiteness makes possible the viewing of practices and subject positions as racialized, as well as racist. Similarly, Richard Dyer, in *White*, stresses that "whiteness needs to be made strange."[49] White is that peculiar sign that is both a colour and not a colour, both visible and invisible. It is that which visibly makes and marks white people as white, while simultaneously working as a signifier of privilege, of what it means to be white, which is invisible (unmarked, unspecific, universal). To make white strange is to uncover the specificity of whiteness, to adapt a wider notion of the white body, of whiteness as involving something that is in but not merely of the body.[50]

Making Noelle white-skinned does not avoid the issue of race, but unwittingly marks the simulator body as unmarked, unspecific and universal. Then again, the moment Noelle leaves for a training session in Afghanistan, the meaning of its whiteness along with its national belonging and dependence on smooth technology shifts significantly. Elias tells the story of Robbie Prepas, a Laguna Beach midwife and consultant to Gaumard, who has experience using various Noelle models in Afghanistan.[51] Afghanistan has, according to the US State Department, the second-highest infant mortality in the world. Evidently, there is some serious work for Noelle to do in this country. In 2004, Prepas was working for

47. Interview: John Eggert, November 22, 2006.
48. Ruth Frankenberg, *White Women – Race Matters: The Social Construction of Whiteness* (Minneapolis, MN: University of Minnesota Press, 1993), 6.
49. Richard Dyer, *White: Essays on Race and Culture* (London: Routledge, 1997), 10.
50. Dyer, *White*.
51. Elias.

the Centers for Disease Control and Prevention to train Afghan medical staff. Prepas and her colleagues used different versions of Noelle, "including one that worked by hand crank to move the mechanical parts. … But while the Noelle mannequins were helpful, power failures and other technological glitches hindered the mannequins' effectiveness. Still, Prepas said Noelle is becoming standard issue in the United States."[52]

The site-specific performativity of Noelle should be emphasized. In the US, the simulator appears to perform flawlessly, but in Kabul its machinery does not work as well, which appears to prevent Noelle from becoming standard issue. It is not clear what technological glitches, apart from the power failures, hindered the effectiveness of Noelle and the delivery of its robot-babies in Central Asia. It appears rather self-evident that white, digitized high-tech bodies would not operate smoothly in a nation into which you must bring a mechanical version that works by hand crank. On the other hand, it is doubtlessly hard to provide a safe, clean environment for childbirths and newborn children in a country such as Afghanistan, fighting to gain some sort of stability in the aftermath of the US invasion and a drawn-out war. If the Noelle project can help save the lives of children and (becoming) mothers, not least in particularly exposed locations, there is a lot to gain by it. As Eggert puts it, "You have to hope for the best, but you have to be prepared for the unexpected. … Simulators can help prepare for the unexpected."[53] Situated in the midst of paradoxical simulatory possibilities and limitations, Noelle not only repeats oppositional relationships between women and machines, East and West, but holds, simultaneously, the possibility of differentiation, of doing things differently.

Death and Simulation

> In 1776, [a] Jaquet-Droz android, a "Musical Lady" that played the harpsichord, was exhibited in London. As she played the five tunes in her repertoire, her eyes would

52. Elias.
53. Interview: John Eggert, November 22, 2006.

> move coyly from side to side, and her bosom would heave lightly, as if she was breathing. The machine was advertised as "a vestal virgin with a heart of steel," but one member of the audience thought her heart might be otherwise.[54]

Female machines are a complex species. Eighteenth century automatons were clockwork constructions, accentuating by their fundamental principles their immortality – as well as the mortality of their makers. A prominent maker of entirely mechanical automatons, predecessors of our modern-day robots, Jaquet-Droz made a range of captivating technobodies. His "Musical Lady" played her instrument dressed in a beautiful, elegantly detailed gown worthy a woman at the French court, crowned by an intricately coiffured wig. Occasionally displayed without her skirt, not only the fine workings of the clockwork machinery were uncovered, but also a pair of shapely legs, covered with delicate silk stockings. The configuration of woman as clock – and clock as woman – is a complex yet manageable creature, mechanical yet lifelike, predictable and pleasing. Jennifer Gonzales points out that in the history of making mechanical bodies, automatons are imbedded in the mechanical innovations connected with keeping time. These are cyborgs that appear more trapped than liberated by their mechanical parts.[55] Immersed in time differently from humans, clockwork bodies mark time, but do not sense the force of time propelling the human body closer to death.

In her *Edison's Eve*, Gaby Wood argues that although androids do not understand death, they are themselves incorporations of mortality: "Rather than being copies of people, androids are more like mementi mori, reminders that, unlike us, they are forever unliving, and yet never dead. They throw the human condition into horrible relief."[56] As such, automatons embody the very antithesis of our mortality, the negative of human knowledge of finitude. What then is the status of death in machines of reproduction, in

54. Gaby Wood, *Edison's Eve: A Magical History of the Quest for Mechanical Life* (New York: Alfred A. Knopf), xiv.
55. Jennifer González, "Envisioning Cyborg Bodies: Notes from Current Research," in *The Gendered Cyborg: A Reader*, ed. Gill Kirkup et al. (London and New York: Routledge, 2000).
56. Wood, *Edison's Eve*, xvii.

machines that reproduce birth, the beginning of life as it were, from which death as the ultimate ending is never far off? Is death present in the Noelle simulation of birth and beginning life? Or, is this machinery rather of the clockwork variety, performing beyond the utmost limit of lived time? If closure is an essential component in the poetics of life, intimately tied to the pleasure of being alive, perhaps the one thing that both prompts and enables us to live, how is closure configured in the simulator? In other words: does Noelle die? A comprehensive answer to these questions must be made elsewhere, as this one is by necessity fairly brief.

Noelle – as mother unit – does not die, unless its machinery breaks down beyond repair. In the manual for S565 Noelle™ (this document has not been updated with the introduction of the S575 Noelle™), there is one passage on death that identifies possible scenarios for becoming mothers, but not hard-coded in the Noelle system: "After delivery the uterus normally contracts reducing postpartum bleeding. Under certain conditions, contraction does not occur and extensive bleeding may continue. If this condition is not recognized and treated in a timely manner the new mother may go into shock and die."[57] However, the newborn Noelle baby, the neonate, will not always make it. In "Teaching Tips" for S565 Noelle™, scenario seven is called "Gloria cord prolapse," which involves a woman, Gloria, and a prolapsed cord emergency: "Prolapsed cord emergencies are life or death situations. This scenario presents a disastrous intrapartum complication that results in foetal death. … Despite heroic efforts of the student team, Gloria's baby fails to recover. The student team must now learn how to comfort the living."[58] As previously mentioned, Laerdal's SimMan can speak of death ("Don't let me die") and also has the ability to die: "SimMan's equipment includes several injured body parts that can be changed depending on the scenario. In the most extreme circumstances, SimMan's internal organs could begin to shut down and he could die."[59]

57. The S575 Noelle Manual is available at http://www.gaumard.com/download/manuals/40_S575.pdf (accessed June 18, 2010).
58. The S565 Noelle "Teaching Tips," accessed June 18, 2010, http://www.gaumard.com/download/manuals/39_S565tips.pdf.
59. Tommy Newton, "A Life or Death Simulation," October 2003, accessed March 29, 2009, http://www.wku.edu/echo/previous/

The fact that the death of the mother is not part of the set-up is intriguing and has several different implications. It clearly sets Noelle apart from the configuration of motherhood, death and the making of artificial bodies in *Frankenstein*. Feminists reading *Frankenstein* have often made sense of Victor's creation of the monster, not as the making of a man by another man, but through the more familiar framework of female sexuality, reproduction and ultimately abortion.[60] Such readings miss the point that the creation of the monster depends on the absence, or even death, of the mother.[61] Frankenstein's reanimation of a dead body in pieces introduces a way of creating life that no longer needs a mother, and actually, all mothers introduced in the narrative end up dead. The lack of a dying-mother scenario for Noelle instead brings the simulator in line with eighteenth century automatons as machines without end. The becoming mother not being able to die may temporarily spare students from traumatic delivery experiences, but simultaneously obliterates an essential component of life and what it means to live, which further adds to the dis-embodiment of this particular technobody.

The absence of death, moreover, makes for an interesting contrast to the conventional ending for machine heroines. From classic examples of female androids of fiction in E. T. A. Hoffman's "The Sandman" (Olympia) and Villiers de l'Isle Adam's *L'Eve future* (Hadaly), to more recent cinematic incarnations such as Ridley Scott's *Bladerunner* (Pris) and Jonathan Mostow's *Terminator 3* (the terminatrix), machine women tend to be rather volatile incorporations of violent sexuality with disastrous implications. The

archive/2003oct/stories/simman.htm.

60. Jane Donawerth, *Frankenstein's Daughters: Women Writing Science Fiction* (Syracuse, NY: Syracuse University Press, 1997); Sandra M. Gilbert and Susan Gubar, *The Madwoman in the Attic: The Woman Writer and the Nineteenth-Century Literary Imagination* (New Haven and London: Yale University Press, 1979); Ellen Moers, *Literary Women* (London: Women's Press, 1977/1978); William Veeder, *Mary Shelley and Frankenstein: The Fate of Androgyny* (Chicago and London: The University of Chicago Press, 1986); Paul Youngquist, "Frankenstein: The Mother, the Daughter, and the Monster." *Philological Quarterly* 70, 3 (1991), 339-359.

61. Margaret Homans, *Bearing the Word: Language and Female Experience in Nineteenth-Century Women's Writing* (Chicago and London: University of Chicago Press, 1986).

fear of machines that become uncontrollable is entwined with the fear of female sexuality that gets out of hand.[62] As manifestations of such masculine anxiety, the female machine must usually "die" at the end of the story. The Noelle simulator is certainly different in significant ways, in neither being a work of fiction nor made for seduction. Yet, this performative machinery is one without end, and as such the very inversion of the regulatory cultural script for female technobodies. I have argued, often and elsewhere, for the posthuman feminist necessity of alternative female machines (ones that, for one thing, do not die at the end of the tale). Noelle doubtlessly produces birth-giving bodies in a selective fashion. Unable to move away from its position flat on its back, it is a prime simulatory performance of female passivity for trans-national trade and training. Then again, this female technobody is quite rare: it is unstoppable and will ultimately outlive, so to speak, its own maker.

62. Andreas Huyssen, "The Vamp and the Machine: Technology and Sexuality in Fritz Lang's 'Metropolis,'" *New German Critique* 24/25 (1981).

"Make my baby" – The DNA spiral further twisted

Karin Wagner

Cynthia Andersen had made up her mind. She wanted to have a child, but she didn't want to be dependent on a partner. And through the progress of science there are now many new opportunities that her parents never had access to. Her gay friends, Greg and Paul, had just become parents through a new method, practiced at the GenoChoice institute at RYT Hospital. She sat down by the computer, ready to go through the web-based process. When reading the letter from the doctor on the start page, her last doubts vanished. Her child would be healthy and possibly also more talented than herself. On the DNA scanning page, there were several options, for single people as well as for couples. Cynthia chose the first one – cloning herself. Now the image of the scanner appeared, the green blinking oval in the centre was obviously the place where she was supposed to place her thumb. Protruding from the scanner were cables leading to a computer. With her left thumb firmly pressed against the screen, she clicked the scan button. The DNA-spirals on the sides were immediately set in motion, and her genetic profile was captured. After a little while she was requested to fill in her name and a password to secure her profile in the database. Not without a certain anticipation, she went on to review her profile. Her heart sank when she saw that if nothing were done, her child would probably be born with some behavioural defects and some diseases. Luckily, with GenoChoice, you were not completely at the mercy of the freaks of nature. She responded to the request: Don't let your child inherit your genetic shortcomings, and clicked the Upgrade button at the bottom at the page. The next page stated the costs of repairing her genes and Cynthia went through the tables several times in agony. What was it worth to have a guaranteed healthy child? Could she afford it? Maybe she could take a loan, because she wouldn't be able look her sick child straight in the face knowing that she could have spared her the illness. She checked all the boxes referring to serious defects and went on to the genetic enhancements. This week there

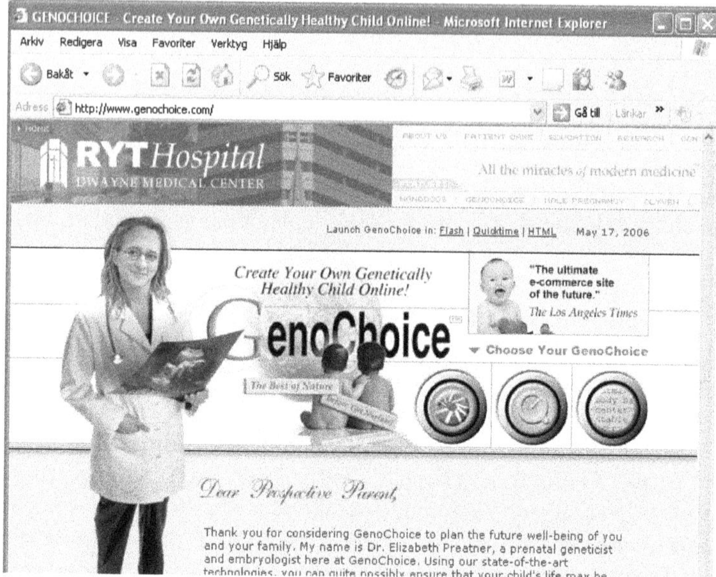

FIG. 1: Start page

was an offer for 50% intelligence increase for only $ 10000. That was an offer Cynthia couldn't refuse, and she also thought that a 50 % increase in artistic talent wouldn't harm her daughter either. She barely dared look at the invoice page, but was in for a good surprise: her insurance would cover all the expenses! She clicked the Make my baby-button without hesitation. The statement from the happy couple on the last page confirmed that she had done the right thing to consult GenoChoice. All she had to do now was to enter her e-mail address and wait for information about the surrogate mother, an image of the embryo and regular updates about the pregnancy. She lent back in her chair and thought of her grandmother–how pleased she would have been if she had lived to know that soon a new enhanced twig would grow on the Andersen family tree.

Does this sound like a parody? It is. In fact, it is a parody of a parody, the web artwork GenoChoice by the American artist Virgil Wong, first launched in 1999. As is often the case in this genre, the parody gives rise to another parody. Whereas my attempt above

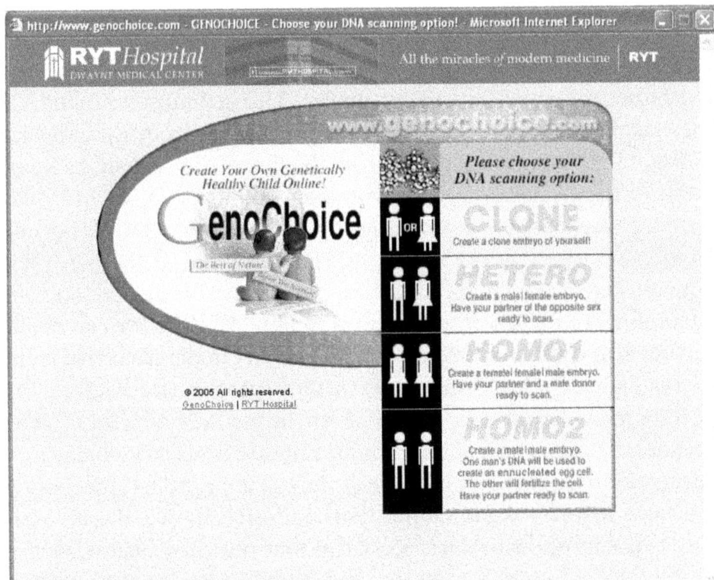

FIG. 2: Scanning options page.

targets a specific work, the source for parody of GenoChoice itself is not a specific text or artwork, but rather the scientific discourse and the marketing discourse around gene technology and reproduction.[1] These discourses manifest themselves i.e. on hospital websites and websites belonging to enterprises in the human reproduction business. Parody is characterised by a comic tone and it can relate to the target in different ways, usually by borrowing the form of it. On the start page GenoChoice (fig. 1), we find the confidence-inspiring image of the doctor and a letter that mocks the persuasive tone medical institutions often use when addressing clients.

In the background of the top banner, the hospital building can be discerned behind the logotype RYT Hospital. These are typical

1. Simon Dentith distinguishes between general and specific parody. The latter targets a specific text or art work, whereas the first concerns a discourse. Simon Dentith, *Parody* (London: Routledge, 2000), 7. His definition of parody is based on function, not on form: "I have defined parody, in a deliberately widely drawn definition, as any cultural practice which makes a relatively polemical allusive imitation of another cultural production or practice." Dentith, *Parody*, 37.

features of a hospital website.[2] Photographs of babies are mandatory elements on the websites of fertility clinics and sperm banks, to evoke the future happiness of their customers.[3] They are mostly smiling babies facing the visitor, like the one in the quote from Los Angeles Times in the upper right corner of the GenoChoice start page. In front of the GenoChoice logotype there are also two seated babies turning their backs at us, looking at a globe. They echo W. Eugene Smith's photograph The walk to paradise garden.[4] This classic photograph depicts two small children walking hand in hand coming out of a tunnel of foliage, from darkness to light. The picture can be interpreted as children facing the future, and so can the seated babies in GenoChoice. In this context an extra dimension can be read into the picture, that the artwork alludes to the future development of gene technology. The two babies form an important visual element that recurs on most pages together with the logotype. So far, the parody has been so low-voiced that it is maybe possible to take the site seriously. The most disturbing part of this first page is the name, GenoChoice. Here a conflict is created between the positive connotations of "choice" as in "freedom of choice" and the title's resemblance to "genocide." The sentence "Create your own genetically healthy child online" sounds like "Manage your bank accounts online" and raises the question of the limits of what could actually be achieved "online." On the following page (fig. 2), the scanning options are described using concepts from scientific discourse such as cloning, embryo and enucleated egg. The parody does not come into full bloom, however, until the scanner page (fig. 3), with the apparatus-like interface and the request to press the thumb against the screen. This is the most comic part of the site, where the parody depends on the artist's invention of a mix between touch screens, optical scanners and DNA-scanners. These devices all exist, but become absurd in this context.

2. As a comparison, see http://www.nyackhospital.org/default.htm (accessed May 5, 2005) and http://www.ivinsonhospital.org/ (accessed May 5, 2005).
3. See for example The Scandinavian Cryobank http://www.scandinaviancryobank.com/ (accessed December 7, 2006) and The Fertility Institutes http://www.fertility-docs.com/index.html (accessed December 7, 2006).
4. The photograph is to be found at for example http://www.masters-of-photography.com/S/smith/smith_children_walking_full.html (accessed May 5, 2005).

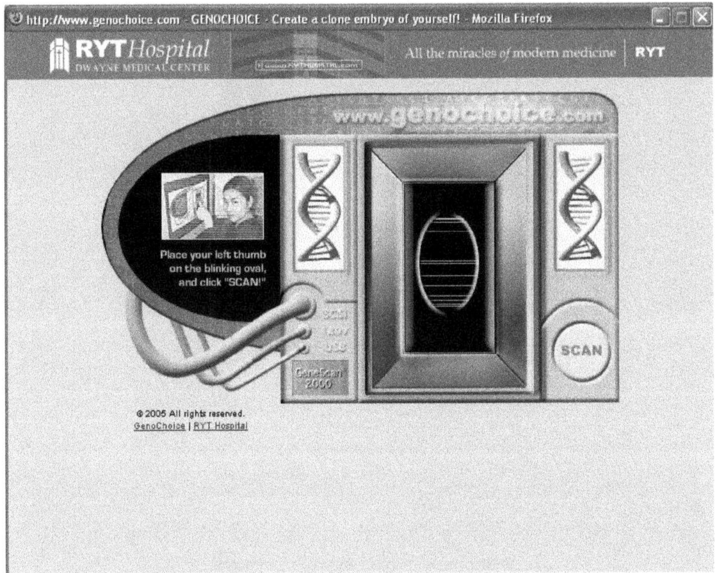

FIG. 3: Scanner page.

On the done-scanning page there is a banner ad linked to a real company. On the next page, the genetic profile page (fig. 4), there is a banner ad that leads to a mock company. With the slogan "Freeze your eggs now!" Eggfrieze Inc. wants to offer their services to us. The "misspelling" of freeze as frieze in the company name is a linguistic marker of parody. The inclusion of the real banner ad increases the parodic effect of the fake one. The tables of behavioral defects and diseases on the genetic profile and the upgrade pages copy the presentation form common for scientific data, and the exactitude of the figures add to the parody of this form. The most conspicuous item in the list of behavioral defects on the profile page is homosexuality. It has the highest probability of all the items and it also appears in most of the scannings I have done.

At the same time same-sex parents are put on an equal footing with traditional parents when it comes to choosing scanning method. Homosexual parents are not likely to "upgrade" their child to heterosexuality. The placing of homosexuality under the heading of "behavioral defects" is a parodic device that has the

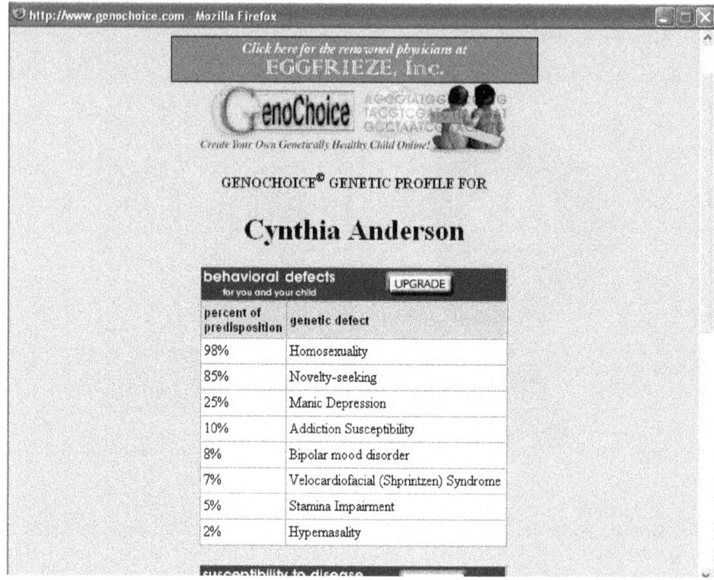

FIG. 4: Genetic profile page.

effect of calling attention to homophobia and to the debate about the causes of homosexuality.

The parody of marketing discourse comes to the fore on the upgrade page. Here we find a time limited discount: This week only, 50% off intelligence and memory enhancement! In order to make the client proceed to the invoice page, production start of the baby is promised within a week. The regular pregnancy updates via e-mail, promised on the last page (fig. 5), functions as a surrogate for the real goods, and offers the client something while waiting for the final delivery. All these features are typical sales tricks to persuade a prospective customer to take the decisive step. While the tone and the form of this last page closely resembles an actual commercial web page, the mentioning of "the revolutionary nine weeks accelerated pregnancy program" is a parodic element that stands out and eventually punctures the credibility of the whole site.

Parody in the new medium of the Internet has many similarities with literary parodies. An extra dimension is added by the process of going through the website procedure step by step and the physical

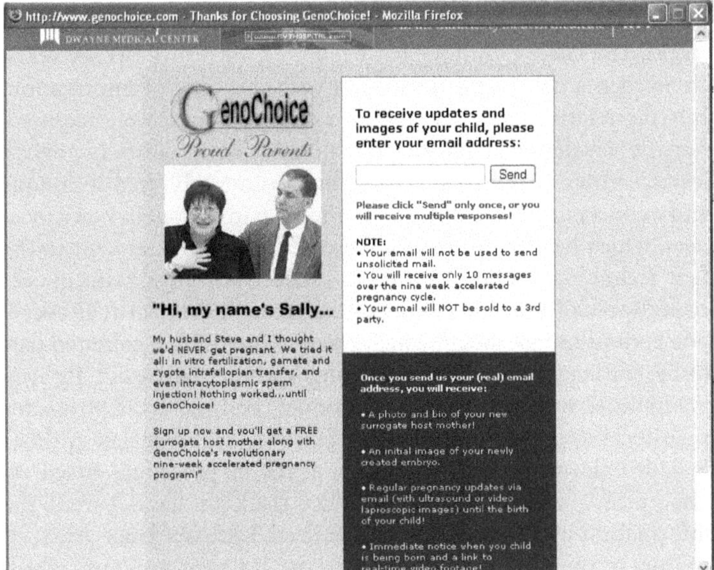

FIG. 5: Final page.

interaction with the screen. When pressing her thumb against the screen, the visitor is made to participate in the enactment of parody.

Identity Through Impression

The fingerprint as an identification method has a long history. In India, in 1877, William Herschel, a civil servant of the British colonial power, started to apply an old Bengali custom, to mark documents with fingerprints. This was a practical way to identify the inhabitants of the area, but the method was not scientifically grounded. The anthropologist Francis Galton proved at the end of the century that each individual has unique fingerprints and that they remain constant throughout life. Thereby fingerprints appeared to be a reliable way to identify people. At the turn of the century 1800-1900 fingerprint files were kept in order to identify criminals.[5] The technique is still in use and at crime scene investi-

5. Carlo Ginzburg, *Myths, Emblems, Clues,* trans. John and Anne C. Tedeschi (Baltimore, Md.: Johns Hopkins University Press, 1989), 122–123.

gations the police nowadays look for both fingerprints and DNA traces. The latter technique has been used for nearly twenty years. In Sweden a file is kept since 1999 with this kind of information. Initially relative large amounts of sample material were required, but the development has made it possible to successfully analyse lesser and lesser samples. Fingerprint is an established technique and a concept that stands for identification in the public consciousness, which has lead to the transfer of the term fingerprint to the new technology. A DNA-profile is called DNA fingerprint in everyday speech.[6] Olfactory research has recently shown that everyone emits a unique smell, which can be visualised in coloured patterns, consequently called colour fingerprints.[7] The search for new possibilities for identification continues, and the areas of use for such techniques are extended. In Sweden, Scandinavian Airlines installed the world's first baggage check in procedure based on fingerprints in December 2006. When the flight has departed, the information will be erased from the local database. Since 2004, all visitors to the US have to submit to the US Visit program, which entails the taking of a digital photographs and the scanning of both index fingers in digital, inkless scanner. The US has also introduced a new passport standard, e-passports, which must contain both biographic information and a digital photograph of the holder.[8] The EU goes even further by requiring fingerprints to be included in new passports for European citizens.[9] This increase of the security level is seen by many as an invasive procedure that infringes on the personal integrity of the traveler. Fingerprint files have up till now belonged to the domain of crime and being included in such a file will for many people amount to being regarded as criminals to be. Maybe this age old technique still awaits its full exploitation and

6. A commission appointed by the Ministry of Justice in Sweden about the extended use of DNA in forensic investigations has the title "Genetiska fingeravtryck" (Genetic fingerprints) Ds 2004:35, http://www.regeringen.se/sb/d/108/a/27189 (accessed May 24, 2005).
7. Smell-seeing, website of the Suslick Research group, http://www.scs.uiuc.edu/suslick/smellseeing.html. Neal A. Rakow and Kenneth S. Suslick, "A Colorimetric Sensor Array for Odour Visualization," *Nature* 406 (2000): 710–714.
8. Crossing US Borders, http://www.dhs.gov/xtrvlsec/crossingborders/.
9. New, secure biometric passports in the EU: http://europa.eu/rapid/pressReleasesAction.do?reference=IP/06/872&format=HTML&aged=0&language=EN&guiLanguage=en.

that state surveillance will rise to a new, Orwellian level with the help of biometrics, all in the name of security.

When the visitor of GenoChoice is requested to place her thumb on the screen, she is supposed to leave a genetic fingerprint. As a basis for a DNA-analysis it will be insufficient, but a genuine fingerprint is actually made on the glass. Is it her most personal and sensitive information that the visitor has exposed? In what databases will it end up? What guarantee does she have that the information will not be misused? Such questions the visitor is likely to ask herself. The Alternative Museum kept an archive with visitors' comments a couple of years after the exhibition. One person wrote that he regretted having scanned his DNA and now wished to be erased from the database.

In the real world, there are online genome databases, but DNA databases linked to individuals exist only in the forensic realm. In the domain of science fiction, to which GenoChoice is close, other rules apply. The workforce of the aerospace corporation in Andrew Niccol's film Gattaca (1997) has to go through a DNA scanning procedure every time they enter their workplace. If their blood sample does not match the record in the database, they are denied entrance. The main theme of the film is eugenics and candidates for space travels are selected according to their genes. But as in GenoChoice the decisions about the offspring is still in the hands of the parents. In a lugubrious scenario of the future, one could imagine the state breeding desirable citizens by means of surrogacy, without even asking or notifying the parents, whose genetic information they have extracted from the database.[10]

Even though it is possible to identify an individual with the help of gene technology, one might ask to what extent our identity is contained in our genes, and if identity can be expressed as abstract information.

Corporality, Information and Reproduction

The American literary theorist N. Katherine Hayles describes in her essay "The Condition of Virtuality" how information theory,

10. See Aldous Huxley, *Brave New World* (London: Chatto & Windus, 1932).

cybernetics and molecular biology developed in the period after the Second World War. Characteristic for these disciplines was the idea that the human body and the information about it can be regarded as separate entities. These entities stand in a hierarchical relationship to each other, where the information has the upper hand. The information theory of Claude Shannon, originally intended for restricted technical application, was in Norbert Wiener's visions of the future transformed to the thought that it would be feasible to telegraph the pattern of a human being from one place to another. Hayles pursues the thesis that all information is dependant on its material basis, and that the embodiment of information is an important aspect that cannot be neglected.[11] If bodies could be described and transferred as information, then the next step would be to reproduce bodies from that same information, and that brings us to a *GenoChoice* scenario. Here the connection between the body and its reproduction has been broken. The link from the parents' to the child rests on information. A surrogate mother is still required, but the bodily union of two persons of opposite sex is a thing of the past. The foetus is sent away to grow somewhere else and the parents observe it at a distance. During the middle ages and several hundred years onwards, it was a custom among the European aristocracy to engage wet-nurses and send children away for a good upbringing and education. The idea that the conception and pregnancy would take place without any bodily involvement of the parents only existed in myths, such as the myth of the stork delivering babies.

Religion has exerted a great influence on our view of sexuality and reproduction. In the Christian mythology procreation without any bodily union is presented as the ideal. Jesus was born through The Immaculate Conception, and Mary's Virgin Birth made her free from sin. GenoChoice can be interpreted as a way to break free from the connection between sexuality and sin, while at the same time the work eliminates "the laws of nature." Seen in

11. Hayles develops this line of thought further in her book *How We Became Posthuman: Virtual Bodies in Cybernetics, Literature and Informatics* (Chicago, Ill.: Univiversity of Chicago Press, 1999). See also Eugene Thacker, "Data Made Flesh. Biotechnology and the Discourse of the Posthuman," *Cultural Critique* 53, (2003): 72–97.

this context the Virgin Mary can be regarded as the first surrogate mother and Jesus as a cloned son of God.

Parenthood in different constellations

What different parent constellations can be found in *GenoChoice*? The single parent, two parents of opposite sex and two same-sex parents. When the visitor has to choose between four different ways to scan her DNA, this means that various forms of parenthood are compared and in some sense placed on an equal footing. Reproduction through DNA information is not an existing method, but there are different forms of artificial fertilisation that *GenoChoice* brings to mind. To clone one self means that one becomes the sole parent of a child that is a copy of the parent. No human being has been cloned as yet, but in 1997 British researchers cloned a sheep. There is an ongoing international debate about whether cloning of human beings should be prohibited by law, and most European nations already have such a legislation.[12] In vitro fertilisation is now a routine procedure for heterosexual couples. Insemination is in some countries available for lesbian couples, which is similar to the Homo1 option, where two women and a male donor are required. But here it is taken one step further; whereas insemination requires only one woman and one donor, in *GenoChoice* both the lesbian partners are to contribute their DNA to the child. A male donor seems still to be required, though. For a male couple there is still no possibility for artificial fertilisation, but there is no theoretical objection to the Homo2 option becoming realised. With the help of egg nuclear transfer, a method that involves emptying an egg of its DNA and replacing it with another, two men could have a child together.[13] This procedure is alluded to in the Homo2 option: "One man's DNA will be used to create an enucleated egg cell. The other will fertilize the cell." This means that no female egg donator is necessary.

While many of the existing fertilisation methods leave one

12. "Kloning" (Cloning), http://www.forskning.se/servlet/GetDoc?meta_id=3829#10. See also Gregory Stock, *Redesigning Humans: Our Inevitable Genetic Future* (Boston: Houghton Mifflin, 2002).
13. Theresa Pinto Sherer, "Can two men make a baby? Researchers say it's possible, but lawmakers must pave the way," accessed May 5, 2005, http://archive.salon.com/mwt/feature/2001/01/31/eggs/print.html.

parent without a genetic tie to the child, the mother in the case of egg donation and the father in the case of sperm donation, GenoChoice involves both parents genetic heritage. In GenoChoice the procreation is always dependent on some male contribution (except in the case of the cloning of a single woman). The Y-chromosome, which only men possess, was thought necessary for procreation, but in 2004 Asian researchers managed to breed a mouse with two mothers and no father with a completely new method. This development shows that sexual reproduction, which builds on a combination of genetic material and permits more flexible adaptation to a changing environment than asexual reproduction, can take new and unexpected forms.

All the options in GenoChoice are dependent on a surrogate mother, a key person who is only mentioned on the last page of the work. Her role seems to be transitory, like a hotbed from where the plants are taken as soon as they are big enough to grow in the garden, i.e. to manage with the "real," genetic, parents. Agreements between these and the surrogate are regulated in contracts, but difficulties can arise when the premises for the transaction are interpreted differently by the parties. In a case study performed in a fertility clinic, Charis Thompson, describes how a surrogate mother became "a temporary relative" to a childless couple. During the pregnancy she was taken very good care of by the couple, which she at that time felt closely connected to. She assumed that the contact would continue after the baby was born, but her interest in the wellbeing of the child was seen as an intrusion by the legal parents.[14] The ethical problems of surrogacy have been much debated, and some debaters advocate the legalisation of commercial surrogate motherhood, whereas others regard these kinds of agreements as a commodification of mothers and children. There is a risk that custody disputes will consider the rights of the parents according to the contract, rather than the best interest of the child.[15]

14. Charis Thompson, "Strategic Naturalizing: Kinship in an Infertility Clinic," in *Relative Values: Reconfiguring Kinship Studies*, eds. Sarah Franklin and Susan McKinnon (Durham, NC: Duke University Press, 2001), 175–202.
15. Hugh V. McLachlan and J.K. Swales, "Babies, Child Bearers and Commodification: Anderson, Brazier et al., and the Political Econ-

California is one of the most surrogacy friendly states in the US, where the intended parents are privileged over the surrogate mother.[16] By contrast, the Swedish parental code states that the woman who gives birth to the child is the one who has legal rights to the child. The passing of this law was seen as an efficient way to deter people from resorting to surrogacy. The longing for children is strong enough to disregard the intention of the law, which a recent case shows. A childless couple made a surrogacy agreement with the sister of the man. Both the egg and the sperm came from the couple. When the child was born, the couple took care of the child and the man was acknowledged as the father. After a year the woman applied for adoption, which was granted her. However, the man changed his mind and revoked his consent. The consequences of the Swedish legislation are that the man and his sister are now the legal parents of the child, although it is genetically tied to another woman. This case points to the problem of legislation as a way to put an end to surrogacy.[17]

Genetic aspects of parenthood are as we have seen tightly linked to the social ones.[18] Children can grow up with a single parent, with grandparents, in a collective, with stepparents, foster parents or adoptive parents. Many children live with divorced parents in new extended families, with stepbrothers and sisters who live with one parent and the other in turns.[19] New words like "plast-

omy of Commercial Surrogate Motherhood," *Health Care Analysis* 8 (2000): 1–18; Elizabeth S. Anderson, "Why Commercial Surrogate Motherhood Unethically Commodifies Women and Children: Reply to McLachlan and Swales," *Health Care Analysis* 8 (2000): 19–26. Alastair V. Campbell, "Surrogacy, Rights and Duties: A Partial Commentary," *Health Care Analysis* 8 (2000): 35–40.

16. John K. Ciccarelli and Janice C. Ciccarelli, "The Legal Aspects of Parental Rights in Assisted Reproductive Technology," *Journal of Social Issues* 61 (2005): 127–137.

17. Anna Singer, "Vem är den rätta mamman?" ("Who is the proper mother?") *Svenska Dagbladet*, August 2, 2006.

18. See for example Jeanette Edwards, Public Understanding of Genetics (PUG) (an EU-project) http://www.les1.man.ac.uk/sa/pug/index.htm (accessed May 5, 2005).

19. Kristina Larsson Sjöberg, *Barndom i länkade familjesystem. Om samhörighet och åtskillnad* [Childhood in Linked Family Systems. Connection and Separation] (PhD diss., Örebro University, 2000); Margareta Bäck-Wiklund and Thomas Johansson, eds., *Nätverksfamiljen. Modernitet och mångfald* [The Network Family. Modernity and Diversity] (Stockholm: Natur och kultur, 2003).

pappa" (plastic daddy) and "bonusmormor" (bonus grandmother) have emerged. Plastic signifies something that is not genuine, but "plastpappa" does not have the same negative ring to it as stepfather. More formal language is required in legal and official circumstances. The Swedish National Council on Medical Ethics explains how they use the concepts genetic, biological and social parent:

> By genetic parent we understand a person who by his sperm or egg has contributed to the conception of the child. By biologic parent we understand the woman who goes through the pregnancy and gives birth to the child. (The man cannot be a biological parent by this definition.) By social parent we understand the person who takes care of the upbringing of the child.[20] (My translation).

Here the usage is different from the everyday usage, where both the mother and the father count as biological parents. For the clients of *GenoChoice*, this would mean that they become genetic parents, but only the surrogate mother is to be regarded as the biologic parent. In this way, childbearing has been relegated to biology, while genetics appears to have been severed from biology, existing in its own abstract, scientific and presumably higher realm. The following quote is from a feminist discussion list on the Internet:

> In the thread "Women take care about children and lose their career" there has now started a discussion about equality in parenthood and the phenomenology in men's and women's parenthood – is it equal? Can it be equal? Are the concepts father and mother equivalent and interchangeable? Is genetic parenthood true parenthood? Shall genetic kinship automatically qualify to "equal parenthood"?[21] (My translation).

20. "Assisterad befruktning" (assisted conception), expert opinion from The Swedish National Council on Medical Ethics, April 5, 1995, accessed May 5, 2005, http://www.smer.gov.se/svenska/yttranden/befruktning.htm.
21. From a contribution to the discussion list Feminetik by the alias Krysset, May 5, 2005. http://feminetik.se/diskutera.

It is exactly these types of questions that *GenoChoice* raises. A reasonable interpretation of the way the four scanning options are presented is that they can be regarded as equal. The heterosexual alternative is not placed first and is not brought forward as the only valid one. An important issue within the gay movement has long been to oppose the hetero-normative ideology of forms of living together. One demand is the recognition of same-sex marriages, but more radical voices claim that this does not promote a fundamental change of the cohabitation forms, that should allow several constellations.[22] Worth noticing is therefore the visual emphasis put in *GenoChoice* of parents as a couple. On the last page of the site, two proud parents, a man and a woman, testify to their satisfaction with the *GenoChoice* program. On the scanning options page, all alternatives are shown with two icons. (In the cloning case there is indeed an "or" between the icons, to indicate that it concerns a man or a woman.) Also the seated babies of the logotype form a couple. On this page there are five images of couples in total, which underlines the concept of pair visually. In the explanation for Homo1 a male donor is mentioned, but his role is more of a catalyst, not an actual parent – the image of the parents shows two women. The traditional constellation mother-father-child is being challenged by these scanning options that in spite of the queer message are conveyed through the conventional symbols for gender – men in trousers, women in skirts.[23]

php?fid=3&mid=69158.

22. Judith Butler points out that marriage as an institution means putting the power of recognition in the hands of the state, which can impinge on the freedom for people to form other types of partnerships. "Is Kinship Always Already Heterosexual?," *differences: A Journal of Feminist Cultural Studies* 13, (2002): 14–44. In her online article "Queer Parents: An Oxymoron? Or just Moronic?", Stephanie Schroeder sees a danger in becoming more normalised and accepted by society by being a parent, and thereby losing something of one's queer identity, http://www.technodyke.com/mamadyke/oxyparents.asp (accessed April 20, 2006).
23. The Austrian sociologist Otto Neurath developed in the 1920s the Isotype system (International System of Typographic Picture Education), on the basis of which international airport symbols etc. were made. An other example of how these signs are used to discuss gender issues is the poster for the film *Transamerica* (2005) that shows the main character, who is transsexual, standing between two doors with the conventional gender restroom signs.

Normality, Outsiderness and Commodity Economy

Francis Galton who introduced the fingerprint as a unique method of identification is also originator of the eugenics concept. According to this doctrine, the human race should be refined through encouraging talented people to breed, while less capable individuals are dissuaded from reproducing. In *GenoChoice* all parents, provided they have the economical means, can take responsibility for the amelioration of the race by upgrading presumably negative dispositions. The lawyer Jennifer Fitzgerald discusses the consequences of biotechnology for people with disability in her article "The Human Genome Project and the Commodification of Self."[24] According to her there is a great risk that they will be even more marginalised by society, through selective abortions, limited access to medical resources and restricted insurance options. Normality will become an even narrower concept than it is today. As a danger of a greater degree she sees the change of the public consciousness that will follow upon the equating of genetic information with human identity. People will not be seen as whole persons, but will be reduced to their DNA-code, especially their deficient genes. As long as it is not possible to cure the illnesses that can be diagnosed by the help of gene technology, there is a risk that biological determinism will prevail. Fitzgerald further criticises science for being intimately connected to capitalist ideology. A decision to give birth to a disabled child will thus be regarded as an irrational economic choice that will burden society at large, not just the parties closest concerned. She cites other researchers, such as Herman Meininger, who has said that parenthood today is more about "child management" than about personal relations.[25]

Fitzgerald writes that just as king Midas in the Greek myth turned everything he touched into gold, Western capitalism con-

24. Jennifer Fitzgerald, "The Human Genome Project and the Commodification of Self," *Issues in Law & Medicine* 14 (1998), accessed May 5, 2005, http://www.metafuture.org/articlesbycolleagues/JenniferFitzgerald/Geneticizing%20Disability.htm.
25. Herman Meininger, "Parental Decisions After Prenatal Diagnosis of Foetal Defects" (paper presented at the 10th World Congress of the International Association for the Scientific Study of Intellectual Disabilities, Helsinki, Finland, July 8, 1996).

verts all it comes near to merchandise. Such a parable is also appropriate for GenoChoice. Parents can order a child by mail order. The child will then be produced, delivered and invoiced. In this process GenoChoice can be seen as both the marketing phase and the launching of the production phase. The association to the upgrade of a computer or a computer program places man on par with a machine that can be enhanced and repaired. All the "gene repairs" that eliminate the risk of defects mentioned in the genetic profile come at a price – and it is up to the client to decide which ones are worth it. As mentioned earlier, GenoChoice makes use of current sales strategies as campaign prices etc. Some business lines sell their main product at a low cost or give it away, while profit is gained on accessories and upgrades. In the GenoChoice case the baby is free and the upgrades of the genetic profile generate the company's incomes. Already at the start page the visitor gets a hint about the theme of the commodification of procreation through the quote from the Los Angeles Times, where GenoChoice is called "The ultimate e-commerce site of the future." This might seem an extreme development, but money is already involved in current methods such as IVF and insemination. Adoptions entail great costs for parents as well.

Web Art and Parody

Art on the Internet often uses parody and appropriation as strategies. Good examples of this are Chelsea and Jonah Peretti's *Black people love us* which parodies racial prejudices through personal homepages, Alexei Shulgin's *Electroboutique* which parodies e-commerce and the hype about electronic gadgets and Susan Collin's *Tate in Space* which parodies the pretensions of art institutions and the discourse around space science. *Tate in Space* was created in 2002 for Tate Online. Like *GenoChoice*, it opens with a confidence-inspiring letter signed by an authority, in this case the director of national programmes Sandy Nairne. In the letter he explains how proud Tate is to be able to show contemporary art in space by sending out a satellite. Collins asked a space scientist to state the exact data for the satellite's position, including a feigned launching date and an orbit, and at what points in time that it would be visible from different places on earth. These scientific data, like the exact percentages in the genetic profile of *GenoChoice*, might lull people into believing that a satellite really exists. The artist sees the driving force behind the visitors of *Tate in Space* belief in the projects feasibility is their wish for it to be true.[26] In the case of *GenoChoice* the greatest driving force to accept it as a current possibility is the fear of childlessness and the fear of having a disabled child.

Sonya Rapoport created the interactive piece The Transgenic Bagel already in 1993. It was later transformed to a web art piece in 1995.[27] In the work, Noah's Ark is supposed to have been the first gene pool, and by choosing a person from the Book of Genesis, the participant can acquire a trait connected to an animal e.g. by selecting Lot's wife, you will be inquisitive through extracting the inquisitive gene out of a carp. By selecting Isaac, you will have endurance like him through the stamina gene extracted from a camel. Then the gene will be processed in three steps and finally impregnated into the bagel, which you have to eat if you want to receive the new trait. In this parody, biblical mythology and gene technology are juxtaposed in the making of sandwiches. When the bagel has been prepared with the desired gene, the visitor is asked the pertinent question if he will eat the transgenic bagel. This corresponds to pressing the Scan button or the Make my baby-button in GenoChoice. Eating a bagel is a mundane act, which clashes with the high-tech procedure of gene splicing and creates a comic effect. This work can be seen as a forerunner to GenoChoice, as it addresses the visitor as a candidate for change, and is driven by the same logic of choosing and processing.

The (inevitable?) Genetic Future

In the popular science press the latest scientific findings are often presented in a way that fires the imagination and fabulous visions of the future are depicted.[28] On the Popular Science website there

26. Discussion about Tate in Space with Susan Collins and Jemima Rellie, February 20, 2002, www.dshed.net/digest/04/content/week2/tate.pdf (accessed May 5, 2005).
27. Sonya Rapoport, "The Transgenic Bagel", accessed August 2, 2006, http://users.lmi.net/sonyarap/transgenicbagel/index.html.
28. See for example Cecilia Åsberg, *Genetiska föreställningar. Mellan genus och gener i populär/vetenskapens visuella kulturer* [Genetic Imageries: Between Gender and Genes in the Visual Cultures of Popular/Science] (Linköping: Tema genus, Linköpings universitet, 2005).

is a special issue dealing with the future of the body. The author starts by enumerating advanced projects already under development – chips that can be inserted in the brain so we can control machines with our thoughts, lungs manufactured in laboratories and pills that make us smarter and more creative – and then he goes on to speculate about the future:

> What will it be like when medications can make a person more monogamous – or religious – and babies can be brought to term in artificial wombs? Read on to learn scientists' plans for altering everything from your bathroom medicine cabinet to your own brain.[29]

Seen in this context, GenoChoice could be taken seriously, as expressed by this comment from a parent site:

> Parodies the *American* (or is it universal) desire to have the perfect child. Frighteningly authentic - there's a part of you that wonders if it might be serious. Difficult to decide if it's funny or downright offensive - depends on your mood.[30]

This ambivalent quality of the work is in my opinion an advantage, as it stimulates critical thinking about biotechnology. GenoChoice highlights issues of importance to both individual and society. Will human identity be equated with genetic information and will this result in a less human society? The increasing amount and diversity of information required in passports to control the movements of people point in this direction. DNA is the most sophisticated way discovered so far for identifying a person, but if cloning of humans becomes possible and accepted, the role of genetic information as a means to uniquely identify a person will change dramatically, since clones share the same genetic informa-

29. John MacNeill, "The Future of the Body," *Popular Science*, July 2005, accessed August 30, 2005, http://www.popsci.com/popsci/futurebody/6a0d9371b1d75010vgnvcm1000004eecbccdrcrd.html.
30. Mumsnet Web guide, accessed May 5, 2005, http://www.mumsnet.com/webguide/22.html.

tion. Clones will then be the people in society with the most freedom of action, since they will be hard to identify. And what about clones as offspring – will a perfect copy of oneself really be an answer to the quest for the perfect child? I doubt it, because of the static effect it would have on the development of the family tree, illusory as that development might be. In GenoChoice it is possible to enhance the clone, but then it is, strictly speaking, no clone anymore. Will childcare be replaced by child management, or will new family constellation provide equally favourable (or detrimental) conditions for growing up as the traditional family? Existing methods such as IVF and adoption have already introduced commercialization into childbearing, but there is no evidence that these children are cared less for than other children. If the possibility to order tailor-made babies will lead to increased intolerance towards people seen as defective is another matter, and it depends on the success of genetic enhanced procreation.

It could well be that the popularity of such procreation will decline as the first generation grows up and it will be discovered that the environment is equally important for the development of a child as the genetic disposition. The nature vs. nurture debate alluded to by the banner "The Best of Nature Before You Nurture" on the first page of GenoChoice will have to be continued.

In Vivo/ In Silico/ In Vitro: The Death of the Choreographer?

François-Joseph Lapointe
Martine Époque

"There have always been two kinds of original thinkers, those who upon viewing disorder try to create order, and those upon encountering order try to protest it by creating disorder."[1]

1. Introduction

Darwin's theory of natural selection states that different individuals in a population are not equal with respect to reproduction and survival.[2] The fittest individuals, those that are able to reproduce and survive, thus have a selective advantage over the others, as they transmit their genomes to the next generation. This iterative process of differential survival and reproduction is one of the strongest evolutionary forces; it is the cause of biological diversity.[3] However, natural selection and evolution would not be possible without the generation of individual variability.

There are two principal sources of variation among individuals within a population. The primary one is genetic mutation; a natural process that results as entropy increases in the universe. Mutation often occurs during cellular replication and increases the diversity of a population. The second source is sexual reproduction, wherein the variation from mutation is reshuffled then passed on to the next generation; children differ from their parents because they inherit genes at random from both their mother and their father. Survival of the fittest is possible when some individuals are selected over the others: the strongest, the fastest, or the sexiest may reproduce more successfully. If the variation induced by sex and mutation is the fuel of evolution, natural selection is the main force driving it.[4]

1. Edward O. Wilson, *Consilience: The Unity of Knowledge* (New York: Knopf, 1998).
2. Charles Darwin, *On the Origin of Species by Means of Natural Selection* (London: John Murray, 1859).
3. Douglas J. Futuyma, *Evolution* (Sunderland: Sinauer Associates, 2005).
4. Sewall Wright, "Genic and Organismic Selection," *Evolution* 34 (1980): 825-43.

Genetic algorithms (GA) or evolutionary algorithms (EA) are inspired by the processes operating in biological populations to transform computer objects using the principles of natural selection.[5] Practically, any GA has two phases: (1) the generation phase and (2) the selection phase. The generation (or creation) phase aims to transform a population of individuals from one generation to the next through reproduction (sexual or asexual) and mutation. The selection phase then acts as a sieve, through which the fittest individuals in the population (i.e. the parents for the next generation) are chosen using a given selection criterion. Iterative repetition of the generation and selection phases thus mimics the evolution of a biological population of individuals that reproduce and survive according to the selective forces of the environment.[6]

The specific use of GA in art is relatively new and best considered in the context of computer applications to several artistic domains.[7] It is within the fields of music and visual arts that such methods have been the most widely-adopted, probably because it is much easier to manipulate sound and images than movement.[8] Although computers have been frequently used by choreographers to create pieces for human or virtual dancers,[9] there has been limited application of GA or EA to dance.

In a recent paper, we introduced the *Choreogenetics* algorithm to generate variation in movement sequences through genetic mutations and selection.[10] This approach has also been used to create interaction among actual and virtual dancers.[11] In the present es-

5. John R. Koza, *Genetic Programming: On the Programming of Computers by Means of Natural Selection* (Cambridge: The MIT Press, 1992).
6. David E. Goldberg, *Genetic Algorithms in Search, Optimization, and Machine Learning* (Reading: Addison-Wesley, 1989).
7. Christiane Paul, *Digital Art* (London: Thames and Hudson, 2003).
8. Colin G. Johnson and Juan J. Romero Cardalda, "Genetic algorithms in visual art and music," *Leonardo* 35 (2002): 175-84.
9. Tom W. Calvert et al., "Applications of Computers to Dance," *IEEE Computing Society* 25 (2005): 6-12.
10. François-Joseph Lapointe, "Choreogenetics: The Generation of Choreographic Variants Through Genetic Mutations and Selection," in *Workshop Proceedings of the Genetic and Evolutionary Computation Conference*, ed. Franz Rothlauf (New York: ACM Press, 2005), 366-369.
11. François-Joseph Lapointe and Martine Époque, "The Dancing Genome Project: Generation of a Human-Computer Choreography Using a Genetic Algorithm," in *Proceedings of the 13th ACM International Conference on Multimedia*, ed. Zhang et al. (New York: ACM Press, 2005), 555-558.

say, our goal is to demonstrate that movement sequences generated by such computer algorithms can be compatible with other sequences, either created by a human choreographer or by other species, namely bacteria. In other words, we seek to determine whether the choreographic processes operating *in silico* (i.e. within the computer)[12], *in vivo* (i.e. within the human body) or *in vitro* (i.e. within the test tube) can be used jointly to generate a coherent dance piece. Furthermore, we want to use this new form of trans-specific (across species) model to facilitate collaboration among virtual and biological entities, digital and actual choreographers, as well as advanced and primitive life forms.

The following sections will introduce our working hypothesis and then provide details about the choreographic operators required for generating movement sequences *in vivo*, *in silico* and *in vitro*. To do so, we will present some basic concepts in linguistics, genetics, and information theory that will be used to develop quantitative measures to evaluate and compare sequences produced through different choreographic approaches. We will then propose two different experimental tests to verify the prediction from our hypothesis and implement our model in a performance setting. Finally, we will present possible extensions and generalizations of trans-specific collaboration to other artistic domains.

2. The Choreographer is a Selective Mutagen

This essay, at the interface of science and art, relies on the hypothesis and prediction that the human choreographer (or any artist for that matter) is a *selective mutagen*. In other words, if a choreographer can create dance pieces through the modification and selection of movement sequences, we postulate that any other process that can do the same choreographic operations could also generate a dance. Indeed, if the role of the choreographer is to compose a dance through the creation and transformation (mutation) of movement sequences, the generation and mutation phase of a GA can play the same role. Then, if we

12. The first public use of the expression *in silico* was made in 1989 by the mathematician Pedro Miramontes during the workshop "Cellular Automata: Theory and Applications" (Los Alamos, New Mexico) to define biological experiments performed on a computer.

assume that the choreographer will select among possible movement sequences, each of which could potentially be included in the final piece, the selection phase of a GA can do likewise. Under this simplistic (and unrealistic) model of dance composition, the choreographer may be substituted by a GA, which also acts as a selective mutagen[13]. Furthermore, the choreographer may also be replaced by biological species that can transform and select movement sequences, namely bacteria. Is a choreographer always necessary for the creation of a dance? That is the question that Merce Cunningham has been asking for the longest time;[14] that is also the motivation underlying our work. Provided that distinct approaches to dance composition produce similar movement sequences, such different generative processes may be used jointly in a novel type of trans-specific collaboration between the human, the machine, and a primitive life form.

In order to test our hypothesis, we need a coherent system to translate movement sequences into digital objects that can be transformed by a genetic algorithm, and molecular objects that may be transfected and evolved into bacteria. For the sake of coherence, the genetic code appears to be the optimal available system to translate these movement sequences into DNA strings. The following section briefly presents the relationship between this coding system, linguistics and information theory.

3. Linguistics, Genetics and Information Theory

The genetic code is a grammar and DNA is a language that allows every life form to reproduce and evolve (i.e. the code is universal). Like every language, DNA has its own vocabulary, syntax and se-

13. It is very important to point out at this stage that our *selective mutagen* hypothesis is intentionally simplistic to allow for a possible collaboration among different choreographic processes. Not a single human choreographer relies solely on mutation and selection to create a full piece, yet most apply these operations to generate movement sequences. Modelisation is the art of making choices. We present one model of dance composition, one that is easy to test and implement *in silico*. Like any model, it represents a simplified version of reality. But a good model is one that makes an accurate prediction, and experimental evaluation of the selective mutagen hypothesis will be used to test the predictive power of our model.
14. Roger Copeland, *Merce Cunningham: The Modernizing of Modern Dance* (New York: Routledge, 2004).

mantics. The nucleotides of the genetic code (A, T, G, C) are the letters of the alphabet, and the codons are triplets of nucleotides that form the words: the amino acids.[15]

The DNA syntax only recognises 20 such amino acids out of the 64 possible triplets, with many synonymous codons (i.e., the code is degenerate). The combination of codons into sequences of amino acids is what characterizes the genes, the sentences of the DNA language. These sentences are then organised into chapters (the chromosomes) and the combination of chapters forms a book (the genome).[16]

3.1 The Information Content of a Sequence

The application of information theory to the genetic code allows for a measure of the complexity of a DNA sequence, with more complex sequences defined as more informative.[17] Namely, information (H) can be measured as the number of bits required to characterize the complexity of a string of length *n*, or the uncertainty in the prediction of what the next word in a sentence will be:[18]

$$H = - \sum p_i \log_2 p_i \qquad (1)$$

where *pi* is the relative frequency of each word in the sentence. Because these absolute values are not comparable for strings of

15. Sungchul Ji, "The linguistics of DNA: Words, Sentences, Grammar, Phonetics, and Semantics," *Annals of the New York Academy of Sciences* 870 (1999): 411-7.
16. The metaphorical representation of the genetic code as an information system has been discussed by Lily E. Kay in her book *Who wrote the Book of Life? A history of the Genetic Code* (Standford University Press, 2000), but this position has been rejected by Richard Lewontin (*Science*, 16 Feb. 2001), among others. This epistemological debate lies outside the scope of our essay. Suffice to say that any other translation system may be used to encode and decode movement sequences, so long as the same code is used *in vivo*, *in vitro* and *in silico*. Our trans-specific collaboration model and selective mutagen hypothesis can also be generalised to other codes. Yet, it may be of great interest to assess the differences among several translation systems for comparison purposes, using the tests proposed below.
17. Chun Li and Jun Wang, "Relative Entropy of DNA and its Application," *Physica A*, 347 (2005): 465-71.
18. Tomas M. Cover and Joy A. Thomas, *Elements of Information Theory* (New York: John Wiley, 199).

words that represent different vocabularies, relative information can be computed as:

$$H' = H / H_{max} \qquad (2)$$

where,

$$H_{max} = \log_2 V \qquad (3)$$

and V represents the number of words in the vocabulary. Thus, the information content of the DNA string {AAAAA} is 0 bits, because there is no uncertainty in predicting what the next nucleotide of the sequence will be. On the other hand, the information of the DNA string {GAATC} is 0.579, being more complex (heterogeneous) and less predictive than the previous sequence.[19]

Information theory is fundamental for understanding the *open work* concept introduced by Umberto Eco.[20] In an open work, the anticipation of the public is a consequence of its uncertainty. This anticipation relies not only on the prediction of what is expected (order), but mostly on the thrill of the unexpected (the "orderly disorder" of Katherine Hayles[21]). Uncertainty thus accentuates the aesthetic experience. Still, there exists a threshold above which information is undesirable, when uncertainty is maximal, the message is entirely random and pure noise emerges.[22] The precarious balance between compatible order in a message (predictability) and complete disorder (uncertainty) is what characterises aesthetics or beauty (B), also defined in mathematical terms by Birkhoff as B = O/C, where O represents order (symmetry, ho-

19. Sahotra Sarkar, "Decoding 'Coding': Information and DNA," *BioScience* 46 (1996): 857-64.
20. Umberto Eco, *The Open Work* (Cambridge: Harvard University Press, 1989).
21. N. Katherine Hayles, *How We Became Post-Human: Virtual Bodies in Cybernetics, Literature and Informatics* (Chicago: University of Chicago Press, 1999).
22. This interesting trade-off between order and disorder, predictability and uncertainty, is also related to Michel Foucault's position in *L'ordre du discours* (Paris: Gallimard, 1971). The order of the genetic discourses makes no exception. The genetic code is submitted to a very strict grammar and syntax, upon which the survival of an individual depends. DNA repair mechanisms are following these grammatical rules to preserve only functional genes, while deleting superfluous mutations. As such, the cleaning process operating at the molecular level plays the same role as *taboos* in the human discourse.

mogeneity, repetition, equilibrium) and C represents complexity (asymmetry, heterogeneity, randomness, disequilibrium).[23] Such concepts of complexity, order, information and beauty are necessary to compare the movement sequences generated by the different choreographic approaches.[24]

3.2 The Comparison of Sequences

When multiple sequences are available, it is of great interest to determine which sequences are more similar by measuring the distances between them. There exists several indices to compare molecular sequences, and these measures also apply to any text (and movement) sequences. The Levenstein (or string-edit) distance counts the minimum number of operations required to transform one sequence into another, by inserting, deleting and/or substituting words.[25] Practically, this index thus provides information about the relationships among different sequences. For example, the edit-distance between {AAAAA} and {GAATC} is 3 because three substitutions are required to edit the first sequence so that it exactly matches the second one. These distances are also used to assess the convergence rate of the genetic algorithm.

4. The Choreographic Process

The creation of a dance piece is a complex process that involves the creation of movements, the definition of a syntax that combines these movements with one another, and the choice of dancers, costumes, props, lighting and music to accompany the dance. Most of these choices are irrelevant to our hypothesis. Namely, a dance can be performed without any music, specific lighting, costumes

23. George D. Birkhoff, *Aesthetic Measure* (Cambridge: Harvard University, 1933).
24. Although Birkhoff's equation of beauty represents an over-simplification of aesthetic experience, it can be computed from DNA strings and easily implemented in the selection phase of a genetic algorithm. Other criteria such as those proposed by Vilayanur Ramachandran and William Hirstein (*Journal of Consciousness Studies* 6 (1999): 15-51) in visual arts have also been used by Ivar Hagendoorn for dance composition (*Journal of Consciousness Studies* 11 (2004): 79-110). Yet, it is still unclear to us how such neurological criteria for evaluating beauty may be translated into mathematical equations and computer algorithms.
25. Vladimir I. Levenstein, "Binary Codes Capable of Correcting Deletions, Insertions, and Reversals," *Soviet Physics Report* 10 (1966): 707-10.

or even humans if we are dealing with virtual dancers. However, a dance cannot exist without any movement. For the sake of simplicity, we will limit the role of the choreographer solely to the creation and modification of these movements. Furthermore, the movement vocabulary will be fixed *a priori* to control for elements of style, only to focus on the syntax of the choreographic language. Thus, the choreographic process here is restricted purely to the generation and transformation of movement sequences, and the selection among several possible sequences to create a dance piece. This approach will be repeated *in vivo*, *in silico*, and *in vitro* to compare the performance of different "choreographers," each representing one of many different incarnations of our selective mutagen hypothesis.

4.1 *In Vivo Generation of a Choreography*

With no loss of generality, let us assume that a fixed number of dancers are involved in this choreographic process, and that a common vocabulary of movements is defined before hand. Within this controlled environment, the choreographer is then free to play with any number of movements to create a piece using standard choreographic operators to combine the movements into phrases. To do so, various improvisational techniques can be used to generate sequences, and these sequences may be transformed and selected to maximise subjective aesthetical criteria.[26]

THE GENERATION PHASE. There exists a wide array of standard choreographic structures in dance composition, but most can be generated by the same syntaxic operations on movement sequences. Namely, movements can be added or eliminated from a sequence to make it longer or shorter. Parts of a sequence can also be edited through the substitution of some movements by other movements that are selected from the same vocabulary. The repetition of a given movement in the dance can generate a redundant sequence. Similarly, a backward sequence can be obtained through the inversion of the order of movements in a sequence, in whole or in part. When a section of the sequence is copied and moved to

26. Lynne A. Bloom and L. Tarin Chaplin, *The Intimate Act of Choreography* (Pittsburgh: University of Pittsburgh Press, 1989).

a different part of the choreography, a movement displacement is observed, Finally, new movements can be generated in the course of the creative process, and the choreographer then has to decide whether these movement merit selection or not.

The combination and juxtaposition of different sequences performed independently or simultaneously by different dancers is also part of the choreographic process. The composition of a solo, a duet, or a group piece involves additional strategies to transform a single movement sequence. Here the objective is to create interactions among dancers, rather than to modify the individual sequences. For example, multiple voices imply that the dancers will perform different sequences at the same time, whereas unison implies the synchronous execution of the same movement sequences by every dancer.

THE SELECTION PHASE. For a human choreographer, the selection phase is the most important one. It is here that different possible sequences are compared subjectively to select only the best, depending on what "best" may mean. For some choreographers, this process aims to tell a story (a narrative process) or simply to trigger an emotional response in the public. For others, narration is not as important as the language itself, the choreographic operators that link one movement to the next.

Regardless of the choreographer's intentions, a dance is primarily created for performance in front of a public.[27] Consequently, aesthetic criteria are usually involved in the selection phase of the creative process; these employ subjective judgement. However, similar syntaxic operations may generate coherent choreographies, and objective measures of "beauty" (such as those proposed by Birkhoff or Ramachandran and Hirstein) may replace the subjective judgement of a human choreographer in some cases. Thus, any creative process that can filter out the variability among possible mutant sequences may do "as good as" the choreographer. If this is the case, such alternative approaches to dance composition may be combined to generate trans-specific pieces of art.

27. Larry Lavender, "Intentionalism, Anti-Intentionalism, and Aesthetic Inquiry: Implications for the Teaching of Choreography," *Dance Research Journal* 29 (1997): 23-42.

4.2 *In Silico Generation of a Choreography*

The *Choreogenetics* algorithm simulates the process of biological evolution to create mutant movement sequences. The generation phase produces movement sequences and transforms them through different types of genetic mutations. The selection phase then applies an array of selection criteria to assess the fitness of the mutant sequences and to determine which one(s) will be transmitted to the next generation.

THE GENERATION PHASE. Contrary to most genetic algorithms, sexual reproduction is ignored, with mutant sequences reproduced asexually by cloning alone. Following this, the daughter sequences will be identical to the maternal sequences and the sole source of variation is through mutation[28]. Depending on the number of sequences (dancers), different mutations are allowed for either one (simple mutations) or multiple sequences (complex mutations) at a given time.

The first simple mutation is a *substitution*: the replacement of one or more movements from the mother sequence by a new movement selected at random from the common vocabulary. This genetic substitution is akin to the choreographer editing the sequence by replacing one movement by another. An *insertion* is the addition of one or more movements, selected at random from the common vocabulary of the mutant sequence; it is the same operation as a choreographer's addition of a movement within a sequence. The opposite of an insertion is a *deletion*, which amounts to the elimination of one or more movements, selected at random from the mother sequence. A deletion thus reduces the length of the sequence; the same operation is performed by a choreographer to delete irrelevant movements. A *repetition* consists of a random selection of one or more movements from the mother sequence, and a repetition of said selection a given number of times in the

28. The interesting use of female names for the mother (original sequence) and daughter (mutant sequence) is required to conform to the linguistic conventions associated with clonal (asexual) transmission of genetic material, such as that simulated by the genetic algorithm. As such, the genetic process is modelled as the transmission of mitochondrial DNA (only transmitted by the mother in most animal species), as opposed to nuclear DNA (transmitted by both parents through sexual reproduction).

daughter sequence. Repetitions are also used by choreographers to create patterns of redundant movements. A *translocation* is the selection of one or more movements from the mother sequence, followed by the reinsertion of this selection in a random position in the daughter sequence; for a choreographer this is a displacement of movements within a sequence. An *inversion* is a deletion of a random segment of the mother sequence that contains at least two movements, and a subsequent re-insertion of this deletion in reverse order at the same location in the daughter sequence. An inversion is akin to a retrograde movement sequence created by a choreographer.

The complex mutations are genetic operations that combine pairs of sequences, thus allowing for interaction among the dancers and the creation of richer choreographies. A duplication is the copy of a mother sequence into a pair of identical daughter sequences. Once a sequence is duplicated, the different copies can be further mutated, independently or jointly, to create unison. Duplications are required to include a new dancer at a particular point of the choreography. An extinction event is the opposite of a duplication, and it consists of the elimination of a mother sequence, when multiple sequences representing different dancers are available. These operations are required to get dancers off stage. Crossover operations are also defined to combine the movements of multiple sequences. Whereas a horizontal transfer selects a random segment from one sequence then inserts it in a second sequence to create a new daughter sequence, a hybridization event creates a mutant sequence by sampling random segments from a pair of mother sequences to produce a hybrid daughter sequence. Such operations are readily employed by choreographers to establish interactions among dancers, using either a single movement vocabulary or multiple vocabularies.

THE SELECTION PHASE. Once a mother sequence is transformed through mutations in a number of mutant daughter sequences, variation is introduced in the population. Subsequently, selection can operate to determine which is the "best" sequence. To do so, the algorithm evaluates at each iteration the fitness of every mutant and compares them to that of the mother sequence. If one of the daughter sequences has a better fitness than the original sequence, it becomes the new mother sequence for the next iteration of the algorithm, otherwise the mother sequence is selected. Thus, the selection phase of the algorithm greatly depends on the number of mutants generated at each iteration, but also on the selection criterion employed.

Several criteria have been proposed to filter out mutants for choreographic purposes. The first one is based on Kimura's neutral (or pan-neutral) model of molecular evolution, which ignores natural selection entirely.[29] Under this model, daughter sequences will always replace the mother sequences, and the mutations will accumulate in a random fashion with the number of generations. However, alternate criteria can be applied to select mutants based on information measures – for example, the sequence with the highest information content may be selected at each iteration of the algorithm to become the next mother sequence. Thus this process aims to select for uncertainty, and thus to introduce increasing complexity in the movement sequences. Yet complexity (and the corresponding concept of algorithmic complexity[30]) does not always meet aesthetic criteria, so that Birkhoff's equation of beauty may also be used as a selection criterion, which also relates to the concept of algorithmic complexity. Finally, a coevolution criterion can also be employed. This selective model implies that every mutant sequence is compared to a target sequence at each iteration of the algorithm using Levenstein distances. The target sequence can be either fixed *a priori* to direct the evolution of the movement sequences, or be defined with respect to a sister sequence which is evolving independently under the same conditions. A choreography generated under the coevolution model will consist of movement sequences gradually becoming more and more similar to a final target sequence.

4.3 In Vitro Generation of a Choreography

If the choreographer is indeed a selective mutagen, any natural or artificial processes that can generate variation in a movement se-

29. Motoo Kimura, *The Neutral Theory of Molecular Evolution* (Cambridge: Cambridge University Press, 1983).
30. Ming Li and Paul Vitanyi, *An Introduction to Kolmogorov Complexity and its Applications* (New York: Springer-Verlag, 1997).

quence can act as a mutagen, and biological mechanisms that filter out the mutants can play the role of the selective agent. A genetic algorithm clearly does both, but we postulate that bacteria transfected with a genetically coded movement sequence can do the same. Indeed, so-called bio-artists have often used transgenic organisms and artificial selection to create artistic pieces.[31] If bacteria can deal with textual and visual messages, they may also be able to transform movement sequences, which can then be selected.

THE GENERATION PHASE. Briefly, the genetic transfection of bacteria by a sequence of movements proceeds in three different steps. The first and most important step encodes the movement sequence in the form of a genetic sequence, by associating each movement of the vocabulary to a unique combination of nucleotides (e.g. AA, CT, GAG). Then, the corresponding sequence can be synthesised in the lab and transfected into a bacterial strain through a bacteriophage;[32] this essentially inserts the movement sequence into the genome of the bacteria. Finally, mutation of the sequence is insured by the use of different mutagens (e.g., antibiotics, or UV light), which will affect the natural rate of molecular evolution. For each division cycle of the bacteria, more and more mutants of the original movement sequence will thus be generated. It is noteworthy that contrary to human choreographers or genetic algorithms, the vast majority of the mutations occurring in bacteria are simple point mutations corresponding to substitutions, deletions and insertions. However, all complex mutations can be obtained by the cumulative effects of simple mutations.

THE SELECTION PHASE. Once different mutants are available, it is necessary to decide upon the "best" to be selected as the next sequence in the piece. As opposed to the choreographer, who subjectively decides what he or she likes based on aesthetic criteria, or the algorithm, which uses an objective criterion to select the best mutants, bacteria are not able to make choices. Rather, the selective agent is provided by the random effect of different culture media and several antibiotics that may be used. Thus, the

31. Martin Kemp, "The Mona Lisa of Modern Science," *Nature* 421 (2003): 416-20.
32. James J. Bull et al, "Experimental Evolution of Bacteriophage T7," *Evolution* 47 (1993): 993-1007.

evolution of a sequence in a bacteria is undirected and unpredictable. However, because it relies on the same mutations as the genetic algorithm, which also correspond to the syntaxic operations of a choreographer, the resulting movement sequences may share some similarities. One objective of our work is precisely to assess the conditions under which these different approaches will be equivalent and compatible.

5. Assessing the Similarity Among Different Choreographic Approaches

Assuming that different creative processes can generate and select movement sequences, it is of utmost importance to assess whether such different approaches can produce coherent results. In other words, we want to determine the conditions under which humans, computers and bacteria may work together to create a trans-specific collaborative piece of art. To do so, one has to demonstrate that movement sequences generated *in vivo*, *in vitro* and *in silico* are not statistically different, that they share the same syntax and exhibit a similar complexity. If the selective mutagen hypothesis is true, it is expected that two movement sequences generated by the same choreographic process will not be more different than a pair of movement sequences generated by independent choreographic processes. In other words, the difference between two sequences created by a human choreographer should not be more extreme than that between a pair of sequences generated either by a computer or a bacteria. Furthermore, estimates of within-group and among-group variance components should not be statistically different. Two distinct experiments can easily be set up to test the predictions of this hypothesis.

5.1 Within-Group Variance

The first validation of the comparative approach assesses the relative performances of independent processes in the generation and selection of movement sequences. Precisely, we want to evaluate the hypothesis that similar choreographies could be obtained by applying the same approach to different movement sequences. Indeed, if the application of a common syntax sufficient for generating choreographies *in vivo*, *in silico*, or *in vitro*, we expect to obtain movement sequences with similar complexity values through the consistent

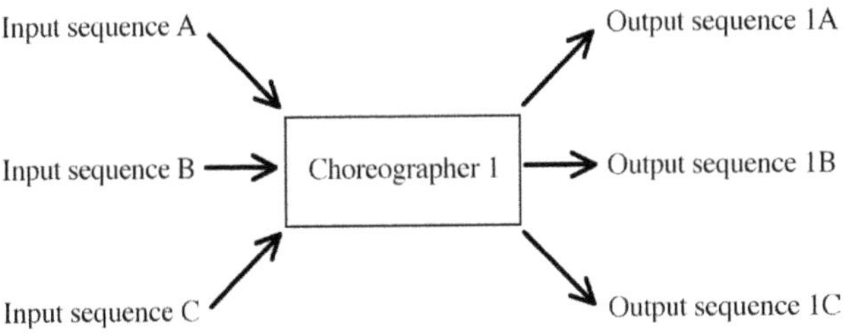

FIG. 1. Evaluation of the selective mutagen hypothesis through the comparison of the within-group variance of different input movement sequences transformed by the same "choreographer."

application of the same model to different mother sequences (Fig. 1). In other words, a choreographer who uses the same basic operations to transform independent movement sequences may generate comparable pieces, from a common vocabulary. Similarly, the same genetic algorithm using a set of fixed parameters should produce movement sequences with similar complexity values.

Testing this hypothesis thus amounts to a comparison of the produced movement sequences by repeating the choreographic process a large number of times under the same conditions, but with independent initialization. For example, the genetic algorithm may be applied to random input sequences to produce mutant sequences under a coevolutionary model. In each case, the complexity of the resulting choreography (or any other measure of aesthetics), is computed to estimate the mean and the variance of the empirical distribution of complexity values. The same exercise may then be replicated with a human choreographer, or with bacterial strain submitted to different mutagens and selective media. If the selective mutagen hypothesis is true, the complexity values computed for the different realizations of the same creative process should be comparable, and the within-group variances should be small. Furthermore, the variance components of different choreographies either generated *in vivo*, *in silico*, or *in vitro*, should be more similar than by chance alone.[33]

33. For all statistical tests involving within and among-group variance comparisons, see Robert R. Sokal and F. James Rohlf, *Biometry: The Principles and Practice of Statistics in Biological Research* (New York: Freeman and Co, 1995).

5.2 *Between-Group Variance*

The next validation of our selective mutagen hypothesis compares the choreographies generated by the different approaches with one another. Whereas the first experimental test aims to compare the performance of the *same choreographic process applied to different input sequences*, the second test focuses instead on the comparison of *different choreographic processes applied to the same input sequence* (Fig. 2). The prediction of the model is that different choreographic approaches using the same basic genetic operators and the same movement vocabulary should generate similar pieces.

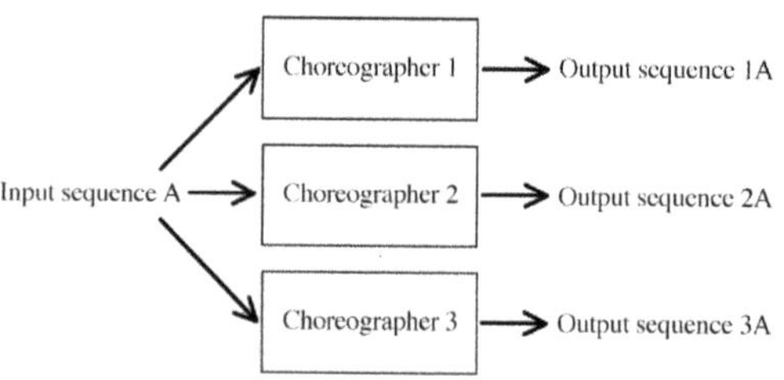

FIG. 2. Evaluation of the selective mutagen hypothesis by comparing the among-group variance of the same input movement sequences transformed by different "choreographers."

Precisely, the test of this hypothesis amounts to the transformation of a given input sequence through mutation and selection *in vivo*, *in silico*, and *in vitro*, and the measurement of the complexity of the resulting choreographies. A large number of repetitions of this parallel process will then establish the empirical distribution of variance components, within-group as well as among-group.

6. Conclusion and Further Studies

Although this paper introduced a simple-minded characterization of the choreographic process, it proposed a precise framework and model to test hypotheses concerning the important role of syntaxic operations in dance composition. We postulated that movement sequences can be generated, transformed and selected through different means. However, further studies are urgently needed to fully investigate the generalization and prediction of our model.

For one, we are currently playing with the idea of mixing different vocabularies into a single choreography to facilitate interactions among dancers through crossover operations[34]. Although our current model only applies to syntaxic mutations, vocabulary mutations are also possible. For example, a retrograde mutation would reverse the order of a single movement, and not the order of the sequence, as would an inversion. Similarly, a mirror-image mutation would flip the orientation of a movement from right to left, or vice versa. Indeed, the addition of new words in the vocabulary by conversion and fusion of contiguous movements has been applied in some of our past experiments. Finally, more complex choreographic forms may emerge by allowing the algorithm or the bacteria to determine not only the movement sequences, but also the position of the dancers on stage, the direction and intensity of the movements, and the tempo of a given section.

Given that the different creative processes are not statistically distinguishable in terms of objective measures of complexity, information or even aesthetics, the next question to address is that of a possible collaboration between these different processes. Obviously, the ultimate verification of our hypothesis will require to collect data on the public responses to such choreographies, to evaluate whether one can actually discriminate among different sections of a dance piece generated *in vivo*, *in silico*, or *in vitro*. Indeed, it is quite possible that two movement sequences with similar complexity will elicit opposite responses (e.g., pleasure and displeasure) in the public. Moreover, trained individuals may perceive differences in the movement sequences that may not be obvious to the untrained eye. The selective mutagen hypothesis is only valid as long as a group of critics (the public) cannot ascertain the difference among sections of a dance piece generated by different "choreographers," or, in other words, that the use of a common set of choreographic operators (genetic mutations) applied to a common vocabulary is a sufficient condition to create aesthetically similar movement sequences, and a coherent choreography.

34. One interesting extension of this work may be to look at other types of grammatic and syntaxic rules, along the language/game metaphors of Ludwig Wittgenstein, *Philosophical Investigations* (Oxford: Blackwell, 1953).

Although our research only addresses dance composition, we strongly believe that the selective mutagen hypothesis also applies to several other artistic fields. We are trying to promote the notion that trans-specific collaboration among digital and biological bodies is already possible, and easily applicable. As long as a coherent model of creativity is developed, a coding system is justified and specific evaluation criteria are defined for a specific domain (e.g., music, visual art, architecture), the generative processes operating *in vivo*, *in vitro* and *in silico* can be used jointly to produce a piece of art. What we present is merely a framework for truly becoming post-human artists.

Biomimetics: Emulation and Propagation in Post-traditional Ecologies

Timothy Weaver

Introduction

The strategy of applied biomimetics has been heralded as design innovation inspired by nature. The inclusive process takes in phases of bioprospecting, biophilia, and emulated biosemiotics to arrive at an implementation of engineered design that ideally might carry implicit biological/environmental sensibilities. Applied elements from this strategy can be found in cybernetics, therapeutics/ medicine, structural engineering, militarized intelligence and now in the propagation of new media art practices. While the idealized practice emulates a platform of honoring/sustaining the original through the preservation of context, the extended practice in post-traditional ecologies frequently crosses the lines of biopiracy and sustains a distancing spiral of simulacra through a re-wiring of ecological consciousness.

This essay will examine the intent and impact of applied biomimetics across a spectrum of creative and technologically-mediated processes through three comparative lenses: as the scientific characterization of natural biological systems; as historically-rooted sociocultural performances/practices and as contemporary engineered-design and creative media-based expression. The objective of comparison through these three perspectives on the biomimetic approach to the appropriation and emulation of natural systems, is to align the phenomena and practice within the contextual realms of traditional and post-traditional ecologies as a means of characterizing the distance between origin and derivative in this creative biocultural practice.

Biomimetic design practice is enabled and delineated by the collective lexical observations of biomaterials/composition, biological time, space, and the enchained interactive patterns of bio-, zoo- and ecosemiotics. In this transference between wild-type ecological contexts and anthropogenic constructs, the thread of intent, distance from origin and resulting impact of application vary according to underwriter, author and user/audience.

The outlined comparisons will highlight these subjectivities in practice and realization while attempting to connect biomimetics as a practical cultural phenomena that emulates, appropriates and propagates across traditional and post-traditional ecological contexts.

Biological Context of Mimicry

In delineating a space for this dialogue on biomimetics, multiple platforms warrant definition as ground for discussion, including: mimesis as a phenomena of biosemiotics within a traditional biological or ecological context and biomimicry as an expressionistic/linguistic pattern with lengthy ethnoecological roots that have become intertwined across origins in animism and bioculturalism.

Mimicry as a biosemiotic phenomena was initially described as a classically characterized ecological phenomena in the 1862 field accounts of H.W. Bates.[1] Bates' Amazonian observations laid down the fundamental eco/zoosemiotic relationship of commensalism in *Lepidoptera*. The biological context of mimicry has since referred to imitations in nature as a survival rationale where the traits of one organism, such as toxicity of a *Danaus* butterfly specie to predators, gives survival coverage to a secondary specie that can adaptably emulate it's phenotypic patterns of expression. In this classic era of Western field biology, extended contributions by Fritz Müller, furthered this documentary platform regarding the mutualistic relationships bridged by mimicry within a biological context.[2]

Batesian and Müllerian mimicry overlap and differ in relation to the reinforcements of natural selection that flow in a biosemiotic stream among models, mimics and predators. Vane-Wright defines this mimetic relationship as involving "an organism (the mimic) which stimulates signal properties of a second living organism (the model), which are perceived as signals of interest of a third living organism (the operator) such that the mimic gains in fitness as a result of the operator identifying it as an example of the model."[3] Bates and Müller's observations took into account visual

1. G. D. H. Carpenter, and E. B. Ford, *Mimicry* (London: Methuen, 1933), 5-19.
2. Carpenter and Ford, *Mimicry*.
3. Richard I. Vane-Wright, "A Unified Classification of Mimetic Resem-

pattern and associated phenotypic emulation as a crossover of bio/zoosemiotic relationship between the model and mimic with the operator's insertion of biophobia as reinforcement for natural selection and the continuance of the mimic to model, or original to derivative relationship.[4]

Mimicry in a biological context also has observation points within networks of social insects. The phenomena of chemical and tactile mimicry within ant colonies is practiced by a variety of opportunistic and symbiotic organisms that become colony guests as a means of benefiting from the activities, interactions and control that delineate the superorganismal construct of ant societies. The colony guests or myrmecophiles, include a diverse variety of flies, beetles, wasps, mites and other insect types that practice some form of mimicry for acceptance into the colony.

As Holldöbler and Wilson discuss: "An ant colony possesses a complex system of communication that enables activities in food gathering, brood care, and other social activities, but makes possible instant recognition of nest mates and discrimination of foreigners. This identification and discrimination system functions like a social immune barrier: only colony members are allowed to enter the ant society, and alien individuals are harshly rejected. Nevertheless, by using various techniques, a considerable number of solitary arthropods have managed to penetrate ant nests. The fact that the ants treat many of these alien guests amicably suggests that the guests have broken the ants' communication and recognition code. In other words, they have attained the ability to speak the ants' language of mechanical and chemical cues."[5]

Wasmannian mimicry was originally postulated from observations of such acceptances of foreign insects into the marching columns of army ants.[6] Wasmann observed the mimicry to be on a tactile level; that is the acceptance into the colony was through light touch via the antennae of ant host workers onto the abdominal petiole of deceptive beetles. This tactile communication in

blances," *Biological Journal of the Linnean Society* 8 (1976): 25-56.

4. Carpenter and Ford, *Mimicry*.
5. Bert Holldöbler and Edward O. Wilson, *The Ants* (Cambridge, MA: Belknap Press of Harvard University, 1990), 471-478.
6. Holldöbler and Wilson, *The Ants*.

combination with chemical mimicry are the key identifiable components of acceptance in this situation of social parasitism. Other cases of symbiotic acceptance of foreigners by social insects involve mimicry in locomotion patterns, behavioral imitation of solicitation signaling, and pheromone mimicry as additional zoosemiotic bridges between the transplanted foreigner and the collective superorganism.

In anthropomorphizing these observations, one is tempted to rationalize the intuitive nature of mimicry as an enchained emergence from insect societies to the human superorganism. Here we have examples of the fundamental observations of the origins of mimicry as an emulated or reproduced thread between zoosemiotic systems in traditional ecologies. The imitation of visual pattern, movement, corporeal form and textures are all examples of phenotypic expression that are advantageously bartered for a leg up on survival and cross species societal acceptance. The questions of intent with respect to this inherent practice cannot be fully probed at this level of entomological observation, but recognition of these phenomena of language cracking and manipulation as fundamental to the biological context of mimicry should be registered in the emergence, persistence and scaling of this phenomena in higher organisms and their superorganismal constructs.

Biomimicry and Bioculturalism

Pre-dating and running concurrent to the previously described scientific record, human ecosemiotic and ethnoecological relationships have been integrated into the lexicon of biocultural emulation and expression throughout the timeline of human history. As lay cultural phenomena, biomimetic integrations pre-date, have co-evolved and provided complementary ethnobiological witnessing to scientific dissections and characterizations. Cultural integration of bio/zoosemiotic reflective performance ritual spans the spectrum of ancient, indigenous, new age and contemporary cultures as emulation of our ethnoecological interactions. Embedded in these emulations is the biophilia hypothesis (the affinity and propagation of the human connection with other lifeforms) that is contiguous with vernacular explorations/exploitations of biologically derived materials, animal behavior, communication, interactions and habitat. Continually emerging from the biophilic response and aesthetic is the emulation of observations as language that continues to be propagated as material and non-material manifestations. The active evolution and lifecycle of these emulations ties expression to the context and sustenance of biocultural memory. The relevance and usage of this root of expression is supported by the interdependence of biological, cultural and linguistic diversity among contemporary global indigenous peoples in Africa, Asia, the Americas and Oceana.[7] Global cross mapping of endemic languages and higher vertebrates highlighted the overlap between linguistic and biodiversity. A global comparison of endemic languages and flowering plant species exposes similar findings.[8] Here we have evidence that the tenets of biological sensibility are feeding the foundational elements of language and expression, with a cross-threaded dependency for persistence.

In broadening the coverage of this work into contemporary Western societies we are now presented with the absorption of ethnoecological relationships into the linguistic ground of the societal frameworks that feed on creative mimicry, interpretation and expression. The spectrum of this framework includes engineered designs, scientific rationalizations and extends into the periphery of what can be described as experimental artistic constructs in pursuit of both practical and innovative manifestations of expression.

The recombination of scientific and vernacular observed biomimetics have previously resulted in aesthetic interventions and creative expression in material culture space. The Arts and Crafts Movement of the 19th Century, and its emulation of ontological motifs in the offset of machine aesthetics, is a grounding example of aesthetic intervention that historically permeated material culture with organic, biological and ecological sensibilities.

7. Luisa Maffi, "On the Interdependence of Biological and Cultural Diversity," in *On Biocultural Diversity*, ed. Luisa Maffi (Washington, D.C.:Smithsonian Institution Press, 2001), 1- 50; Gary P. Nabhan, "Cultural Perceptions of Ecological Interactions," in *On Biocultural Diversity*, ed. Luisa Maffi (Washington, D.C.:Smithsonian Institution Press, 2001), 145-156.
8. D. Harmon, "Losing Species, Losing Language: Connections Between Biological and Linguistic Diversity," *Southwest Journal of Linguistics*, 15 (1995): 89-108.

An Open Source | Bioindustrial Partition

In the contemporary timeframe, a design practice with biomimesis at the emulated core has been firmly established as a process that observes/mimes/mines traditional ecological sources, emulates and technically transcribes, decodes and recodes biosemiotic relationships and re-mediates the reductionist resultant into a post-traditional ecological context. These creative/engineered applications cover a broad subjective spectrum in terms of intent, impact and ethics as they relate to origin and derivative. This creative practice as an idealized process can ethically emulate and honor the source of biomimicry and propagate a realization of environmental awareness while resulting in a material output as intellectual property, product, or rendering. The focus on biomimicry as an idealized practice of environmental consciousness described as "innovation inspired by nature" has been reflected on by Janine Benyus[9] and David Suzuki and is now one of the essential mantles of the Bioneers. Benyus' Biomimicry Guild professes and propagates this process as an environmentally ethical approach to implementation and education of biomimetic design practices.[10] Benyus has fostered a Biomimetics Institute, a business consulting service, a "Conservation for Innovation" program and an open source biomimicry database in collaboration with the Rocky Mountain Institute in following the green mindset of "quieting human cleverness, listening to life's genius, echoing what we learn, and giving thanks to nature."[11] Keeping the biological commons open as a resource for the propagation of innovation, invention, expression and language is essential to this version of biomimetic design practice. Therein the value of nature can be recognized and diversity preserved in a generative cycle of bioculturalism.

Parallel to this philosophy of an open biological commons is the longer standing practice of privitised reductionism that, in the 1970s, was deemed biotechnology. This design practice also intersects the collective record of ethnoecology but has parsed and processed the collective experiential lexicon though a sequence of

9. Janine M. Benyus, *Biomimicry, Innovation Inspired by Nature* (New York: HarperCollins Publishers, 2002), 287-284.
10. Benyus, *Biomimicry*; Biomimicry Guild, http://www.biomimicry.org.
11. Benyus, *Biomimicry*.

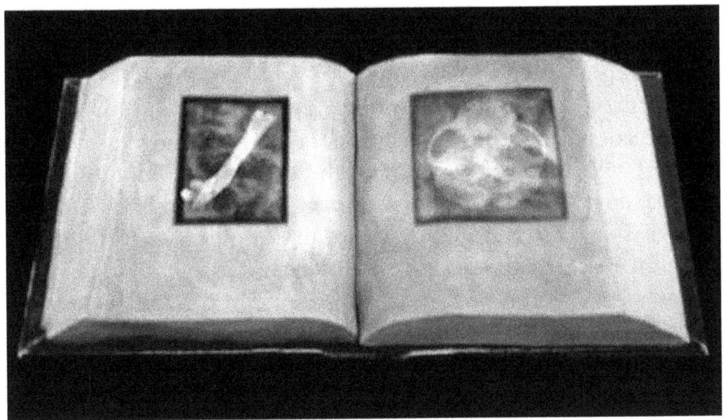

FIG. 1. Timothy Weaver, *Biological Narrative 1 thru 5 (primata)*, 2004.

bioprospecting, bioscreening, bioprocessing, biopatenting and in some cases biopiracy.[12] Biomimicry as design innovation inspired by nature is still at practice here, but the flow back to the commons is shunted and a privitised-form of innovation, expression and language is locked down within the bioindustrial context. A post-traditional ecology ensues but the semiotic connections are also gated by venture capital, intellectual property and investment rules. Without a flow back to the source, derivative expression and language become self-propagating in the vein of simulation spiraling into simulacra, in this case bioculturalism gives way to bioindustrialism. Dissemination and assimilation of biological and environmental sensibilities ride on market lines. Despite the fact that this biomimetic design practice draws from the commons both ecologically and academically, the first person accounts of this context of ethnoecological rewiring is not openly accessible nor conversant for an exchange on the thread of intent nor the impact of this mode of biomimetic design practice on the distance from origin.

In speaking to this point of privitisation and the sequestering of biomimetic design practice, I give testimonial to my own past practice as an industrial microbiologist and environmental

12. Vandana Shiva, *Biopiracy, the Plunder of Nature and Knowledge* (Boston: South End Press, 1997), 1-7.

engineer in a bioindustrial context. The dissemination of results of this design practice as open source is strictly curtailed by the ownership of intellectual property rights that are signed over to the sponsor of your investigations.

In comparing the propagation of language and expression in this privatized context to that of the biological commons, I also must contrast it against my current investigatory practice as a new media artist engaged in an aesthetically-driven form of biomimetic research and design. The biomimetic foundations of one of my recent works, *Biological Narrative 1 thru 5*, 2004, include the concepts of the re-narration of extinction, the emergence of life, biological time, the dilemma of sustenance and the internal drama of pain. The aesthetic mediation of these works employs a poetic remix of video/audio sources filtered through computational routines of artificial life and predator-prey ecological models as captured and interactive digital cinema. The resulting work converges content into a biological narrative – a storyweb that integrates an overlay of multimedia sources onto a bio/ecosemiotic backbone. The modularity of these media works are now resulting in new performative or live cinema works that are using biomimesis as an interpreter for real time media dialogues. My intent in this authoring is to enable an expectation of biological and environmental sensibilities within media forms that are routinely discussed as post-human or post-biological. In producing this work I have been in pursuit of an understanding of the range of intent and inquiry exerted by new media artists who are hybridizing biomimesis into emerging media forms/expressions that can be recognized as biomimetic art.

Biomimetic Art

The propagation of new language, expression and dialogue related to biomimetic design practice has also been manifested in the tangential context of new media art. The manifestation of biomimesis in new media also draws upon the questions of intent with respect to distance from source, process, and bioethics. The extended dialogue also falls back on the implication brought up previously in this essay with respect to the interdependency of bioculturalism, biodiversity and linguistic diversity.

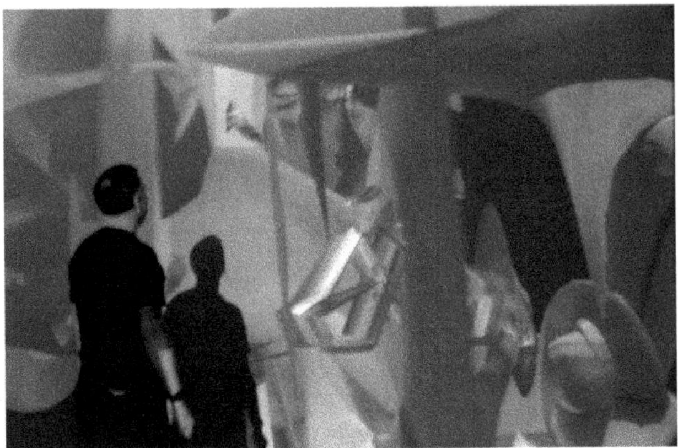

FIG. 2. David Stout, *100 Monkey Garden*, 2005.

In the exploration of this dialogue, three practicing new media artists were interviewed for their first person response to the implicit issues that surface in this genre of creative inquiry. The interviewed artists, David Stout, Gordana Novakovic and Steve DiPaola are all utilising biological modeling/biocomputation in their work for the production of interactive electronic media in an installation or digital cinema format. During the dialogue with these artists, reference was brought up with respect to audience response to the works in tension against real life, particularly in the case of synthetic ecologies being looked upon as simulacra and as a diversion from real time issues of environmental impact and habitat desecration.

In framing a dialogue, the artists were asked to provide 1) a descriptive platform/synopsis of a recent work that they have created in the domain of biomimetics and new media. The follow-up question to this synopsis covered 2) the issues of investigatory intent, distance from origin and the potential resulting impact of the cited project with crossing the lines of what we might relate as biomimetic.

David Stout is a new media artist, Program Coordinator for iARTA (Initiative for Advanced Research in Technology and the Arts) and Professor of Composition in the College of Music at the University of North Texas, Denton Texas. David's current work with artificial life-based generative cinema is predicated on his long-standing interest in creation mythologies, immersive aesthetic experiences and audio-visual structures. David describes his evolving project *100 Monkey Garden* (2005) relative to the domain of biomimetics.[13]

> My previous biomimetic explorations include the creation of contemporary dance dramas depicting fictional biota. These early works were closely aligned to the idea that humans are consummate *copycats* and that the richness of diverse multi-cultural artistic practice can be traced to our innate sense of biomimicry. These ideas closely mirror the theories underlying acoustic ecology, which posits that the evolving history of a culture's musical aesthetic is closely aligned to the soundscape of any given cultural epoch.
>
> The *100 Monkey Garden* is my first attempt to model a self-generating ecosystem based on algorithmic mutation and predator-prey relationships. The work is a significant step beyond the surface emulation of animal-like behaviors often associated with the earliest biomimetic artforms such as, indigenous music and dance rituals, in that it relies on a broad set of behaviors that guide the evolution of form within a complex system. In the development of this project I cannot overstate the importance of "noise" as a *prima materia* in that it provides a wide range of quasi-random behavior within a broad field of frequencies that can be isolated, filtered, harnessed, sculpted and/or distilled to produce divergent sonic, visual, kinetic behaviors. This is an important distinction or addendum to the practice of Noise Art as a whole, which is typically associated with extreme experiences of audio frequency, amplitude and associated visual phenomenon.

13. David Stout, interview by author via e-mail, Santa Fe, NM to Denver, CO, May 17, 2006.

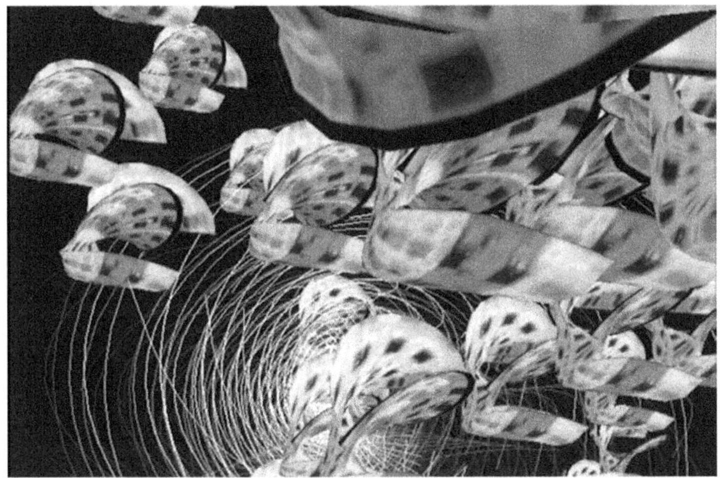

FIG. 3. David Stout, *100 Monkey Garden*, 2005.

The work is built on the premise that each aesthetic element will be responsive to every other element - sound will effect image, image will control sound, one organism will repel another while attracting yet another, new forms are born, older forms are eaten, the system remains in balance unless the predator-form fails to keep pace with the growing population. If the population grows beyond 100 entities there will be an increased likelihood of triggering latent morphogenic mutations and a growing possibility that the whole system could eventually come to a halt in either a cataclysmic computer crash or benign frozen screen. The parallels to our own environmental condition are clearly drawn, while the idiosyncratic nature of the audio-visual forms expand the vocabulary of abstract image making and sound art independent of the biomimetic narrative.

In the design of the work, I have combined a number of isolated digital techniques into a comprehensive systemic vision that reveals increasing levels of connectivity. These techniques or components include, dynamic visualization

> of three dimensional forms, sound synthesis generated by formal motion, audio-visual recursion, simulation of attraction and repulsion between visual forms, independent spatial movement, recombinant procreation, predation, rules governing population dynamics, skin pattern, texture and recombinant genetic history, networked computer system allowing separate computers to drive differing synthetic entities, etc. The successful implementation of these components suggests the possibility of a series of dispersed installation sites networked as a complex system of independently evolving niche ecologies.

David addresses the investigatory intent of this project as it relates to the distance from its biological sources and origin, from the following perspective:

> I work within the uneasy paradox of the exponential growth of computing power to simulate compelling mediated realities while environmental destruction results in the extinction of increasing numbers of the world's species. On one hand, the artist and scientist can take the role of creative deity, whether playing out carefully constructed inquiries or wild postulations through digital genetic simulations that safely avoid the dark implications of wet-lab genetic engineering. On the other hand, the artist can imbue the fictional media eco-system with a socio-political perspective that questions or reinforces our environmental conscience. Considering the growing state of environmental crisis and the potential calamities that could unfold due to genomic interventions, is the simulation of a complex web of connectivity really enough? This is not a simple one answer question but rather a stimulating poke in the ribs to help us clarify the pragmatic resonance of our intentions. One thing is for certain, documentary film and coffee table photography which aestheticise environmental decay and species destruction can only go so far to stimulate positive action before freezing the event as just another fetishised aes-

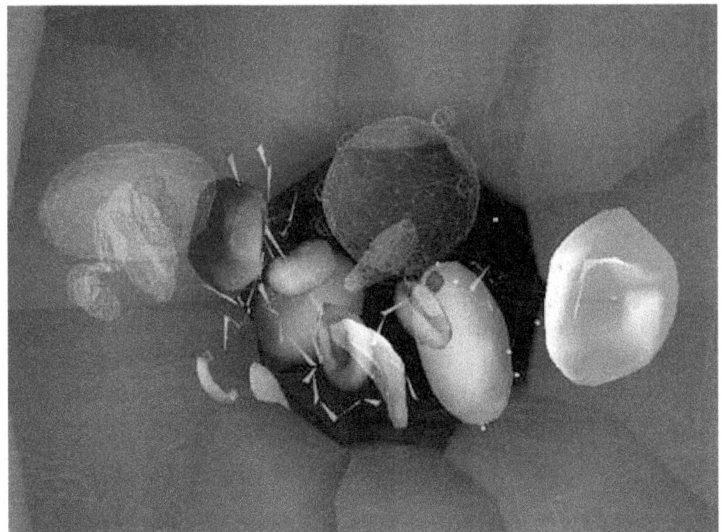

FIG. 4. Gordana Novakovic, *Fugue*, 2005.

thetic object. As artistic mediums, photographic essays and documentary films cannot approach the power of real-time dynamic media to directly engage the viewer-participant in a compelling sensation of interaction with a responsive interconnected world. The crux of this work is not to merely create a new narcotic in the form of an artificial paradise, but rather, to amplify the experience of interconnected participation in something larger than ourselves. This kind of simulation allows us to witness the consequences of our actions (or inactions) to affect unfolding events over many generations.

Gordana Novakovic is Artist-in-residence in the Department of Computer Science, University College London. Gordana describes the platform of her recent art/science collaborations in the investigatory realm of biomimetic art.[14]

14. Gordana Novakovic, interview by author via e-mail, London, England to Denver, CO, May 22, 2006.

> *Fugue* (2005) is a scientifically informed interactive art project based on the functioning of the human immune system. It is inspired by the musical form of fugue, and operates within the framework of an artificial immune system algorithm, expressed through vision and sound. The emergent, evolving nature of the artificial immune system algorithm, the use of repetition in the form of a succession of variations of "events," and the complex structural and functional interrelationships between the individual elements and processes are strongly related to the musical form of counterpoint, which formed one of the inspirations for the artistic concept for *Fugue.* The sound is presented as a mental soundscape, a resonance of the function of the immune system in the body.
>
> What marks *Fugue* is its approach to interactivity – as a way of engaging the participants, as a means of managing the relationships between the constituent elements of the software, visuals and the sound, and as the primary artistic methodology. The participant, immersed in the virtual world of inter-sensory experiences, engaged in a spontaneous non-verbal dialogue with the responsive intelligent medium, establishes a technology-mediated introspective relationship. Although it implements Dr. Peter Bentley's computational model of the immune system, *Fugue* is not a computational anthropomorphic simulation. The focus is on the body of the participant and on the complex processes induced by the specific properties of an interactive immersive environment.

Gordana's perspective on the distance and context of this work in relation to its biological origin follows:

> *Fugue* symbolises the inseparable interconnectedness between all particles and functions of a living body, which is shaped by its inner functions as much as by its interaction with the world. The technology-mediated environment of the contemporary metropolis manifests

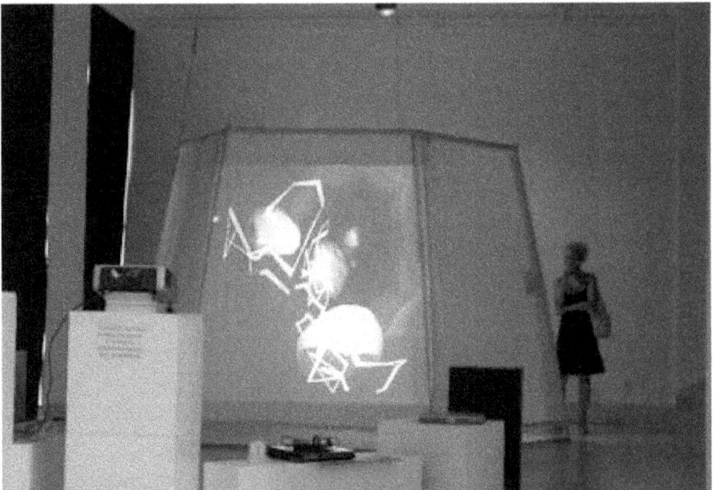

FIG. 5. Gordana Novakovic, *Fugue*, 2005.

interdependence between the human and technology, leading to the experience of multiple realms such as reality/virtuality and presence/telepresence. Further complex changes have occurred in the perception of time/space, noise/silence, speed/duration and movement/stillness. In this context, *Fugue* deals with the emerging issue of the ecology of mixed realities. It aims to address a set of ethical and philosophical concerns such as the interdependence of the human and technology, and the impact of specific technologies on the human body.

The fugal structure helps to achieve one of the major aims, by not only representing the processes involved, but at the same time painting a larger picture of the role of the immune system in the functioning of the human body and mind. This illustrates the immune system's intimate interconnectedness with the total sum of particulars that constitute each human being, and creates a metaphor affirming holism as one of the fundamental principles of transdisciplinarity.

> With regard to biomimesis, the focus is on understanding and applying the principles of biological processes rather than simply creating beautiful imagery, or re-representing scientific findings as visualizations or sonifications. In order to underpin the focus on processes, the conventional "real" images of the lymphatic system have been reduced to symbols, and the roughness of the clay models of cells that were subsequently digitized has been deliberately kept present. The sound, composed by Rainer Linz, has been developed following an analogous methodology.
>
> *Fugue* is designed to challenge the established approach to interactivity, and to explore what happens to the body, and consequently to consciousness, in an audio-visually rich, technologically dense interactive environment. We do not expect to arrive at a definite answer, but rather we expect to shed some light on the issues that we find important not only within the narrow scope of interactive art, but more importantly in everyday life, where our environments, packed with technology, are being converted into interactive cities. We believe that to raise these questions from a non-scientific, but scientifically informed, viewpoint may make a significant contribution to the ecology of perception, emphasising the critical and political aspects of new media art.

Steve DiPaola is a new media artist and scientist who directs the social based Interactive Visualization Lab as Associate Professor, at the School of Interactive Art and Technology, Simon Fraser University in Vancouver, Canada. Steve has been conceptually, technically and aesthetically pivotal in biomimetic projects for the public interaction and interpretation with the beluga whale population in the Vancouver Aquarium. Steve describes the platform of his *Virtual Beluga Project*, 2005 as follows;

> When Vancouver Aquarium visitors are asked what of all things they would you like to see for the live beluga

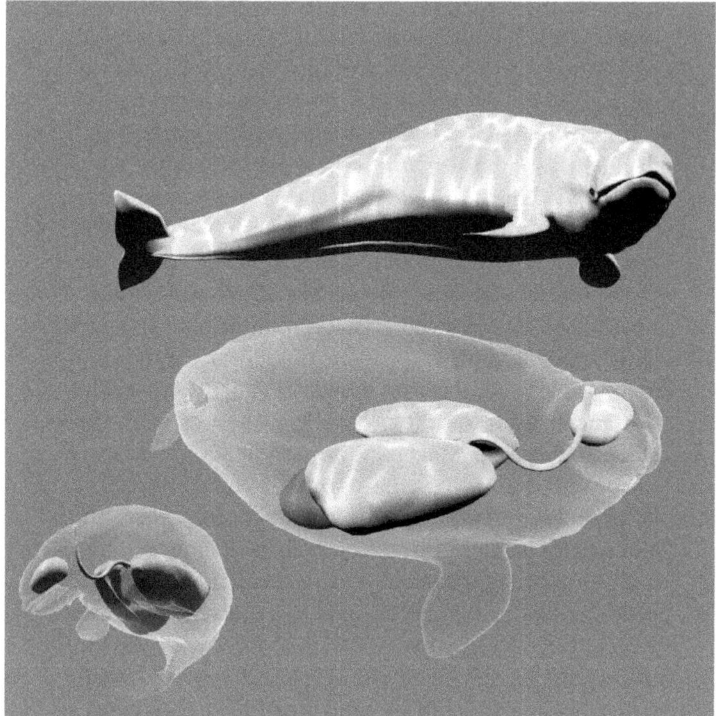

FIG. 6. Steve DiPaola, *Virtual Beluga Project*, 2005.

whale exhibit, overridingly they respond, that they would like to be able to "swim with the whales." When you see these magnificent multi-ton creatures from above the water or through the glass walls, as we humans want to be able to fly with birds, we just want to swim with these animals.[15]

Our work focuses on the design of a real-time interactive simulation exhibit which uses adaptive technologies to better immerse the visitors in complicated concepts about the life of wild belugas. The beluga interactive

15. Steve DiPaola, interview by author via e-mail, Vancouver, B.C, Canada, to Denver, CO, May 22, 2006.

> uses extremely realistic 3D graphics and an action selection system that allows the virtual belugas, in a natural pod context, to learn and alter their behavior based on contextual visitor interaction. The interactive design was informed by: research data from the aquarium's live belugas, interviews with the marine mammal scientists and education staff as well as by its proximity to and integration with the aquarium's live beluga exhibit. Ethogram information of live beluga behaviors was then incorporated into the evolving interactive simulation which uses 3D physically based systems for natural whale locomotion and water, artificial intelligence systems including modified neural networks and an action selection mechanism to simulate real-time natural individual beluga and group behavior. The system allows visitors to engage in what-if scenarios of wild beluga behavior. By giving the simulated 3D whales complex, specific and interdependent behaviors, such as the mother-calf bond or male aggressiveness mixed with age based playfulness, the animals become individuals and their interactions become narratives. Add the visitor interaction and reflected conversation space as visitors interact with the belugas using a shared tangible interface and an additional level of narrative occurs. The domains of biomimetics and new media allow for the group behavior of visitors and whales to swim together for awhile.

Steve addresses the investigatory intent of this project as it relates to the distance from its biological sources and origin, with the following detail:

> In creating this project, many shared ethical, cultural and ontological issues have emerged that have lead to deep questions surrounding virtuality and nature. For instance, we have made a strong effort to ensure the interactive enhances and works with but never replaces the benefit of live belugas, but in doing so; ironically, the virtual belugas represent "wild" nature while the real

belugas can only represent captive nature. By pairing the virtual and the real do lay persons appreciate the idea of nature more than nature itself? To illustrate this point, an example that happened among biology professionals is recounted as follows. We needed to educate a large group of aquarium trainers, educators and scientists on what was possible with the simulation software. As we discussed the possibilities of the memory capabilities of the simulated brains of the virtual 3D whales allowing for new experiences to be learned and retained over the life of the exhibit, a whale trainer got excited that we could have visitors teach the whales typical training behaviors that they could retain for future visits. This brought a sharp and loud retort from one scientist who exclaimed, "No, we won't treat these whales liked trained monkeys too!" as she pounded her hand on the table. It became clear that even trained animal professionals would have trouble with virtual life. For while it is admirable to protect the virtual whales from human intervention and keep them naturalistically pure, saving the purity for the mirror of the whale rather than the real whales has several implications. It brings the point home even more when you realize that the aquariums real whales which live in captive quarters were the source of our data that created our pure and wild virtual whales. Should real animals be poked and prodded to get good sensor data to create perfectly wild virtual animals? While the Vancouver Aquarium takes very good care of the animals, this is not true of all zoos and aquaria. A near future scenario that is still wrought with ethical and culture issues but tries to deal with this issue of treating the real animals better that the virtual ones can be imagined. One could imagine, future zoos and aquaria that house only simulated animals with strong visitor interaction capabilities. These simulated animals would be derived from sensor data (real-time or stored) from live wild animals. The animals stay free and wild while human interactions with their virtual personas become more complex and

> experiential. Again there are serious implications to such a scenario but one we have to face whether it is with interacting with "wild animals" or having a distanced communication between our children and their grandparents in Florida who are holographically projected in real-time on the living room couch.

The artists' comments from above are a sampling of first person critical dialogue, perspective and intent relative to biomimetic design processes that have emerged as new media art. The described biomimetic art projects were built upon a biophilic foundation of ecological/ecosemiotic, immunological/biosemiotic and ethological/zoosemiotic observations and modeling for the development and realization of interactive, interpretive, and generative works through multimedia output. Underlying the technical mediation of these works is the voiced intent by each of the surveyed artists to deliver content, resonate with source and expand dialogue around issues that are at the core of our intellectual engagement with the contemporary issues of ecosystems, corporeality and cross species understanding. The applied processes of biomimetic design for these artists has a mission apart from those in the engineered design field but distinctly fall within a delineated domain which draws on natural sources as a platform for emulation.

In critically assessing these works in relation to their distance from origin to derivative, the perspective and immersive experience of audience must be acknowledged as well as the generative nature of the works. Each of these artists has expressed their intent of uncoupling the enchainment of simulation to simulacra by adapting the narrative structure of their work to honor/sustain/inform the original thru the preservation of memory and context. The narrative structure of the works also allows for a presence of parallel or interdispersed access to reference to origins/source. The configuration of Stout's *100 Monkey Garden* (2005) enables the audience to return to their real world environment for reference while maintaining the opportunity to revisit the ongoing evolutionary state of the work. In Novakovic's *Fugue* (2005) the audience has their own bodies as a constant reference to the computational immunology of the work. In DiPaola's *Virtual Beluga Project*

(2005) an actual pod of beluga whales is comparatively available to the virtual work. The audience perspective on the distance from origin of these works is implicitly challenged by the constant output of dynamic multimedia patterns that are generated from the biological computational sources/models that comprise the mimetic core of each of these works. Because each of these works are generative and therefore open-ended in nature, new ground with regard for the distance from origin comes to the fore as individualised/original content that audience/participants/users can immersively experience as an extended emulation of life systems. Although the potential for simulation and simulacra's diversion exists here, a continuum back to an affinity to biological source fills out the experiential spectrum from memory of origin and on to a horizon of novel pattern recognition.

Moving Toward a Critical Framework for Biomimetics

In framing a dialogue around these works there is perhaps the beginning of a structure for looking at the manifestations of biomimesis as it relates to traditional and post-traditional ecological contexts.

Implicit to the observations, analysis and implementation of the biomimetic design process is a regard for the biophilic aesthetic as an attraction for life processes and characteristics. In materialized expression, interaction and interpretation, biomimesis is actually located on a continuum from the biophilic to the biophobic - that is an affinity and adherence to biological origin and the distancing from these sources to derivative. In this essay the manipulations and expressions of biomimesis have been reviewed from a classic biological context, to integrations in bioculturalism, the appropriations in the bioindustrial world and explorations in what may be described as biomimetic art. With this range of contexts in mind, perhaps we can begin to recognize the essential continuum that biomimetic pathways are reliant upon in their exchange between the affinity for life sources and their distance to derivative.

An Art Still Struggling to be Born: Second Nature, Mutability, Hybridity

Rolf Hughes

The human race in its marvellous and varied works
seems to reveal itself as a second nature in this world.
Leonardo de Vinci[1]

We are used to a certain classification of things.
With the language, or languages, it has become for us a
second nature.
Ludwig Wittgenstein[2]

The history of technology is dated by the existence at each stage of a particular type of machine; the history of the sciences is now reaching a point, in all its branches, where every scientific theory can be taken as a machine rather than a structure, which relates it to the order of ideology. Every machine is the negation, the destroyer by incorporation (almost to the point of excretion), of the machine it replaces. And it is potentially in a similar relationship to the machine that will take its place.
Felix Guattari[3]

The problem of seeking to separate and number categories of nature is that one remains within the logic of sequences, and thus origins, which presupposes that ordering itself is more than mere convenience. A term such as "nature" cleaves the world. Once such a conceptual scalpel is introduced into our reasoning, we devote a great deal of time to looking for things to cut, slice, and dice – and then arrange in a pleasing order. The moment we become aware of its presence, we become aware of its

1. Leonardo da Vinci, *I Manoscritti e I Disegni di Leonardo da Vinci, Il Codice Arundel*, Reale Commissione Vinciana, 4 vols, (Rome 1923–30), 15IV.
2. Ludwig Wittgenstein, *Remarks on the Philosophy of Psychology*, Vol II, (Oxford: Basil Blackwell, 1980), 678.
3. Felix Guattari, *Molecular Revolution: Psychiatry and Politics*, trans. Rosemary Sheed (Middlesex: Penguin Books, 1984), 112.

absence. Hence Schelling's complaint in his *Philosophical Investigations of the Essence of Human Freedom* (1809) that all modern European philosophy, since Descartes, is tainted by the fact that "nature does not exist for it and that it lacks a living ground [*dass die Naure für sie nicht vorhanden ist, und dass es ihr am lebendigen Grunde fehlt*]."[4] Nature, ignored by philosophical chatter, grows *unspoken*. To be unheeded is to be arid, or wild. Others, however, believe that the great intellectual breakthrough of the centuries before Ovid belongs to Aristotle, and specifically his central contribution to western culture which was, according to Caroline Walker Bynun, to address "generation and corruption – coming-to-be and passing away – as not mere fluctuations of appearance, the adding and subtracting of qualities or "skins," but the replacement of one existing substance by another. Real change."[5] Such statements exemplify the dichotomy between articulating a *philosophical* approach to humankind – an essentially *ahistorical* mindset – or else a time-bound, Darwinian (and not infrequently reductive) *historical* narrative, which will be bound by logics of sequences, selections, orderings, and thus origins. To mark out the terrain by planting seemingly hostile species (such as the reassuring continuities of *essence* versus the transformative potential of *metamorphosis*, *phusis* versus *authorship*, *auto-poiesis* versus *design*) seems less a strategy for cultivating productive hybridity than a tangle of incompatibility.[6] Or is it rather the combative logic of that connective preposition "versus" that creates the dissonance? If so, a term such as "second nature" becomes more promising, engirdling as it does both the temporal and the ontological.

4. Schelling, *Philosophische Untersuchungen über das Wesen der menschlichen Freiheit und die damit zusammenhängenden Gegenstände*, in *Sämtliche Werke* (Stuttgart and Augsburg: J. G. Cotta'scher Verlag, 1860), I/7, 356.
5. Caroline Walker Bynun *Metamorphosis and Identity* (New York: Zone Books, 2001), 177.
6. In his attempt to distinguish the craft object from the work of art, Heidegger seeks to describe what lies hidden behind the Greek concept of *techne* as a mode of knowing based on seeing what is present as such in its "unconcealed" nature, a "bringing forth of the being" of something. Thus the artist brings forth the appearance of something (or its physical reality); this is an example of *phusis* (spontaneous growth), which in turn is what distinguishes art from craft.

Alternatively, we might seek a different concept that can bind (or indeed contrast) such oppositions – *agency*, for example. On the one hand, if we are mere animality (*zoé* in Agamben's paraphrase – or "the simple fact of living common to all living beings (animals, men, or gods)"), we are as *un-free* (and as non-creative) as animals.[7] But if our sense of agency (as a defining characteristic of being human) is merely a rationalization of the impersonal cultural and biological forces surging through us, then the concept fails to elevate us above the level of beasts. Agency thus provides a conceptual lens for examining such distinctions as those between human, animal and mechanical, author and afflatus, origin and network. And yet we remain as perplexed as ever as to what it means to be endowed with consciousness – as Raymond Tallis remarks, "we can comprehend how the light comes into the brain, but not how the gaze looks out of it."[8]

It is a striking feature of post-war western culture that while our technologies have displayed increasing capabilities to connect us across time and space, our intellectual discourses have simultaneously emphasised fragmentation, cultural anxiety, and loss of faith in the grand narratives of modernism. Yet the increased border traffic between disciplines, and the increased visibility of digital technologies to artists and scholars, have contributed to an increased interest in interdisciplinary methodologies. As Noah Wardrip-Fruin and Nick Montfort write:

> New media's biggest breakthroughs haven't come by simply expending huge resources to tackle well-understood problems. They have come from moments of realization: that a problem others haven't solved is being formulated in the wrong way, or that a technology has a radically different possible use than its current one, or that the metaphors and structures of one community of practice could combine with the products of another to

7. Giorgio Agamben, *Homo Sacer: Sovereign Power and Bare Life,* trans. Daniel Heller-Roazen (Stanford: Stanford University Press, 1998), 1.
8. In "The Ardent Atheist", interview with Raymond Tallis, Guardian Newspaper 29 April 2006, accessed June 8, 2007, http://books.guardian.co.uk/review/story/0,,1762901,00.html.

> create a third. That is, breakthroughs have come from thinking across disciplines, from rethinking one area of inquiry with tools and methodologies gained from another – whether in the direction of Ted Nelson's conception of computing in literary terms, or the opposite movement of Raymond Queneau's formulation of storytelling and poetry in algorithmic terms. One of these brought us the web; the other, digital narrative.[9]

Bolter and Grusin's concept of "remediation" – "a defining characteristic" of the new digital media, they claim – which they describe as "the representation of one medium in another", reminds us that historically new media have always "borrowed", "repurposed", "reused", "reappropriated", and essentially "remediated" older media forms.[10] This raises the question to what extent practitioners *repeat* existing conventions, or how far they define their practice *against* such conventions (and thereby try to escape what the literary critic Howard Bloom has termed "the anxiety of influence" in his book of the same name).

Walter Benjamin claims that art can anticipate technological developments by creating a demand that could be fully satisfied only subsequently:

> The history of every art form shows critical epochs in which a certain art form aspires to effects which could be fully obtained only with a changed technical standard, that is to say, in a new art form.[11]

Dadaism, for example, "attempted to create by pictorial – and literary – means the effects which the public today seeks in the film." Let us here recall Bynun's summary of Aristotle's major contribution to Western philosophy as the insight into "generation and corrup-

9. Noah Wardrip-Fruin and Nick Montfort, "*The New Media Reader*, A User's Manual," in *The New Media Reader*, ed. Noah Wardrip-Fruin and Nick Montfort (Cambridge, MA.:MIT Press, 2003), xii.
10. Jay David Bolter and Richard Grusin, *Remediation: Understanding New Media* (Cambridge, Mass: MIT Press, 1999), 45.
11. Walter Benjamin, "The Work of Art in the Age of Mechanical Reproduction," in *Illuminations*, trans. Harry Zohn (London: Fontana, 1992), 230.

tion – coming-to-be and passing away – as not mere fluctuations of appearance, the adding and subtracting of qualities or "skins," but the replacement of one existing substance by another."[12] Artifice, in art as in religion, exposes established oppositions such as those between the authentic and the fake, integrity and falsehood, history and fiction, rationality and emotion (this is what Plato called the "circumstantial and mutable effects of human artistry" in regards the pantomimic or mimetic arts). What emerges from the dissolution of such oppositions is a paradigm shift, from the binary logic of the oppositional (the logic of ordering) to the multiple logic of the hybrid. Historically, the avant-garde, according to Mathew Biro, "anticipates the contemporary moment through its practices of imagining new forms of non-bourgeois, hybrid identity."[13] In the dada art of Raoul Hausmann, for example, the cyborg appears as an emblem of modern hybrid identity. Hausmann used the cyborg to represent the part-organic, part-mechanized human, a figure through which images of identity and difference, self and other, could be juxtaposed yet remain unresolved. This necessitated the formation of a new spectator – "an audience whose growing desire for the play of meaning was to result in the creation of new modes of perceiving and acting in the world."[14] In images such as his collage and photomontage *Self-Portrait of the Dadasoph* (1920) Hausmann's notion of the cyborg as a dialectical figure – a hinge between self and other, artist and spectator – anticipates, according to Biro, Norbert Wiener's cybernetic conception of humans

12. Caroline Walker Bynun, *Metamorphosis and Identity* (New York: Zone Books, 2001), 177
13. Matthew Biro, "Raoul Hausmann's Revolutionary Media: Dada Performance, Photomontage And The Cyborg," *Art History* 30 (2007), 26–56: 27. Acknowledging the critique of the terms 'hybrid' and 'hybridity' by post-colonial theory for implying 'pure', original states, generally of a racial, sexual or national nature, Biro intends "hybrid identity" as denoting "a cybernetic conception of the subject". He writes, "This conception of human identity defines it as a process in which different forms of self-identification or allegiance encounter and transform one another without any one of them becoming dominant. In its more radical forms, which appear in some, but not all, of Hausmann's productions, hybrid identity defines not only 'race', 'gender', 'class' and 'nation' to be non-binding, non-essential properties, but also acknowledges the broader interconnections that human beings share with animals, machines and their environments." Ibid, 52.
14. Ibid, 36.

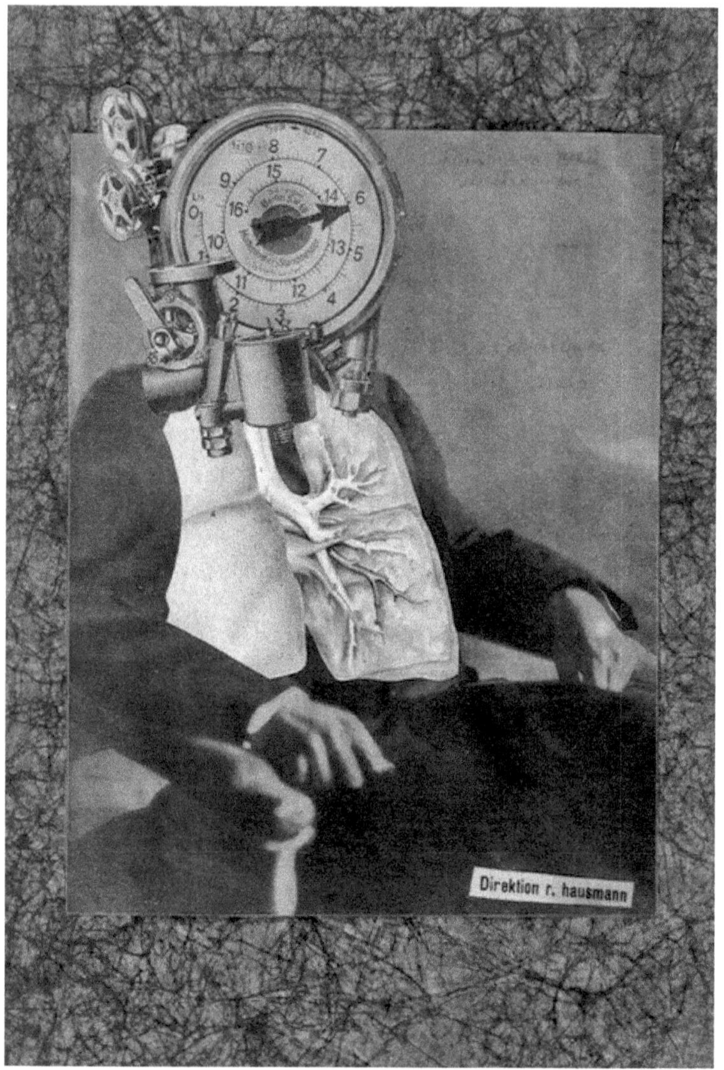

FIG. 1: Raoul Hausmann, *Self-Portrait of the Dadasoph* (1920).

and machines as essentially performative – gatherers, manipulators and producers of information, whose actions and activities were programmed by messages received, while also anticipating the prosthetic experiments of Stelarc in our own day. "What exactly is the economy of prosthesis in the Portrait of the Dadasoph?" asks Cornelius Borck.

> It shows a partly technological, and partly biological or surgical reconfiguration of a human body. The head has been replaced by an ensemble of technology. A combination of pressure gauge and film projector pass for face and brain, whereas the chest offers a look inside the body in the form of an anatomical preparation of the lung with its tubes and arteries. According to the socio-historical reading, this photomontage portrays a cyborg *avant la lettre* with nothing in its head but machines.[15]

Donna Haraway has described the process by which the 'integrity' or 'sincerity' of the Western self gives way to decision procedures and expert systems; "No objects, spaces or bodies are sacred in themselves; any component can be interfaced with any other if the proper standard, the proper code, can be constructed for processing signals in a common language."[16] Such cyborgs represent for Harraway the emancipatory potential of a non-coercive, hybrid identity – "a kind of disassembled and reassembled, postmodern collective and personal self". Difference, here, is not merely tolerated but welcomed – it is, as Biro comments, "an understanding that does not insist on a common origin or nature, let alone a master narrative or theory."[17] In this conception, the cyborg rep-

15. Cornelius Borck, "Sound work and visionary prosthetics: artistic experiments in Raoul Hausmann," *Papers of Surrealism* Issue 4 winter (2005), accessed April 14, 2008, http://www.surrealismcentre.ac.uk/papersofsurrealism/journal4/acrobat%20files/Borckpdf.pdf.
16. Donna Harraway, *Simians, Cyborgs, Women*. Cited in Matthew Biro, "The New Man as Cyborg: Figures of Technology in Weimar Visual Culture". *New German Critique*. Nr. 62, Spring-Summer 1994, 50.
17. Donna Harraway, cited in Mathew Biro "Raoul Hausmann's Revolutionary Media," 26–56. Biro goes on to suggest parallels between Raoul Hausmann's hybrid identity politics and the racially hybrid figures that appear in the photomontages of Hannah Höch throughout the 1920s and early 1930s, the ethnic hybidity later taken up in the

resents a creature of pure information, subject to continuous dispersal, transformation, exchange, its DNA *potentiality* rather than selfhood.

Such assemblages of physical and non-physical matter are characteristically conceived and articulated from within the metaphorical logic of the prevailing technology. For Descartes, the mechanisms of the early seventeenth century that would help him explain how organisms operate were essentially spring-operated and hydraulic automata. As a result, Descartes' thinking was influenced by clocks, watches, water mills, and church organs. Georges Canguilhem writes:

> [A]s long as the concept of the human and animal body is inextricably 'tied' to the machine [because it provides the machine with its power] it is not possible to offer an explanation of the body in terms of the machine. Historically, it was not possible to conceive of such an explanation until the day that human ingenuity created mechanical devices that not only imitated organic movements…but also required no human intervention to set them going.[18]

Bioart, involving the deployment of the latest biotechnology to present living organisms as art, extends such hardware and wetware couplings via DNA sampling, organ transplants, cloning, cryonics, prosthetic enhancement, gene manipulation, and the like.[19]

photomontage practices of Romare Beardon in the 1960s and 1970s and in those of Wangechi Mutu since 2001. The subversion of gender identity, Biro continues, is also characteristic of dada and surrealist assemblage of the 1920s and 1930s, as well as the 'corporeal sculpture' of such contemporary artists as Louise Bourgeois, rona pondick, Robert Gober, Kiki Smith, Charles Ray, and Jake and Dinos Chapman. See Biro, 51.

18. Georges Canguilhem, "Machine and Organism," trans. Mark Cohen and Randall Cherry, in *Incorporations*, ed. Jonathan Crary and Sanford Kwinter (New York: Zone Books, 1992), 49.
19. See for example Eduardo Kac, *Telepresence & Bio Art: Networking Humans, Rabbits, and Robots* (Ann Arbor: University of Michigan Press, 2005); Eduardo Kac (ed.), *Signs of Life: Bio Art and Beyond* (Cambridge Massachusetts and London, England: MIT Press, 2007); and Eugene Thacker, *Biomedia* (Minneapolis: University of Minnesota Press, 2005). For a discussion of how Natalie Jeremijenko's work problematises distinctions

Common to this approach is a view of "life" as information – a design material – rather than as essence, identity or vitality. Eduardo Kac creates what he calls "transgenic artwork that explores the intricate relationship between biology, belief systems, information technology, dialogical interaction, ethics, and the Internet."[20] Stelarc's PARTIAL HEAD digitally transplanted a scan of the artist's face over that of a hominid skull, constructing a 'third' face, one that becomes, in the words of the artist, "post-hominid and pre-human in form". The data was used to print a scaffold of ABSi thermal plastic, via a 3D printer, which was in turn seeded with living cells – becoming "a partial portrait of the artist, which was partially living".[21]Adam Zaretsky's recent project "Mutate or Die" (with Tony Allard) adapts William Burrough's cut-up technique to Burrough's DNA itself. The process is described as follows:

> 1: Take a glob of William S. Burroughs' preserved shit
> 2: Isolate the DNA with a kit
> 3: Make, many, many copies of the DNA we extract
> 4: Soak the DNA in gold dust
> 5: Load the DNA dust into a genegun (a modified air pistol)
> 6: Fire the DNA dust into a mix of fresh sperm, blood and shit
> 7: Call the genetically modified mix of blood, shit, and sperm a living bioart, a new media paint, a living cut-up literary device and/or a mutant sculpture.

This results, the artists claim, in a "new media" described variously as "a living paint [...] scatological clay brought into material being [...] a construction of an infectious metagenomic ballistic shot in the dark."[22]

between the arts and science, with particular reference to the project "Sperm Economy", see José van Dijck, "After the 'Two Cultures': Toward a '(Multi)cultural' Practice of Science Communication."*Science Communication*, Vol. 25 No. 2, December 2003: 177-190.

20. Kac's description of his work "Genesis," accessed June 7, 2007, http://www.ekac.org/transgenicindex.html.
21. From Stelarc's web site, accessed June 8, 2007, http://www.stelarc.va.com.au/partialhead/index.html.
22. They conclude, "We can call it whatever we want, but it is out

If Zaretsky's characteristic hyperbole carries the familiar whiff of boys goofing with their new toys, others explore synthetic fusions with a less scatological focus. Debbie Chachra, a materials scientist at the Olin College of Engineering, Boston, studies natural plastic – so called "bee plastic". This plastic is resistant to biodegrading and, being produced by a species of bee native to New England, is made without the use of fossil fuels and therefore may one day become a source of natural, non-oil-based plastics, replacing part of the global fossil fuel industry.[23] Liam Young and Darryl Chen investigate the cultural consequences of emerging biological and technological futures via future urban scenarios located in the projective worlds of speculation and fiction which betray, in the words of Geoff Manaugh, an interest in "the murky borders between the synthetic and the geological, the organic and the mass-produced."[24] The visionary architect Rachel Armstrong pursues "living architecture" to explore "cutting-edge, sustainable technologies by developing metabolic materials in an experimental setting." In the process, existing distinctions between the living and the inert elements of our urban environment become potentially obsolete. Armstrong comments, "These materials possess some of the properties of living systems and couple artificial structures to natural ones in the anticipation that our buildings will undergo an 'origins of life' style transition from inert to living matter and become part of the biosphere. By generating metabolic materials it is hoped that cities will be able to replace the energy they draw from the environment, respond to the needs of their populations and eventually become regarded as alive in the same

of control," accessed February 8, 2011, http://hplusmagazine.com/2011/02/06/mutate-or-die-a-w-s-burroughs-biotechnological-bestiary/.

23. Accessed February 7, 2011, http://www.olin.edu/faculty_staff/bios/bio_dchachra.html#.

24. Noting that the images of their project "Postcards from a Green Future" are "almost farcically green" ("it's sustainability redone as Grand Guignol"), Manaugh asks, "What if those verdant fields of green out there are actually cloned and genetically-modified? What if that well-trimmed nature is simply an exhibition on display?" From http://www.worldchanging.com/archives/010561.html, accessed February 7, 2011. Liam Young and Darryl Chen's website is "Tomorrow's Thoughts Today," accessed February 7, 2011, http://www.tomorrowsthoughtstoday.com.

way that we think about parks or gardens. Since metabolic materials are made from terrestrial chemistry they are not exclusive to First World countries and have the potential to transform urban environments worldwide."[25]

In the dissolution of distinctions between animal, machine and human, we confront the necessity of rethinking the opposition between *materialism* and *immaterialism*. This opposition is the core of art and religion, and depends upon a theory of representation. In religion, the radical theologian Don Cupitt argues, "being" and "sign" (or "icon") may be false distinctions; a god cannot be separated from the image of wood or stone that was venerated in worship and therefore has never existed as a superior referent, he claims. He makes the following analogy:

> Each and every Donald Duck image published by the Disney Studios really *is* Donald Duck himself; there is no superior original. Donald Duck is a vivid character to millions, maybe billions, but he simply doesn't need to have any existence outside his own iconography. It would be a pedantic mistake to try to establish the existence of a real Donald Duck independent of the standard image, and then to investigate whether the standard Donald Duck image is in fact an accurate likeness. No, the vitality and cultural influence of Donald Duck does not depend at all on any such question. It depends entirely on the vitality of the image and the way it behaves. And because signs are infinitely multipliable, and each of them is the real thing, Donald Duck can be omnipresent. So it is with the god. His being but just a sign, and as the case of cartoon characters and figures such as Uncle Sam and Santa Claus shows, it is perfectly possible and indeed very easy for someone who is only a sign to be a vivid and influential personality known to everybody. Every culture has dozens, perhaps hundreds, of such figures.[26]

25. Rachel Armstrong's website, accessed January 23, 2011, http://www.rachelarmstrong.me.
26. Don Cupitt, *After God: The Future of Religion* (New York: Basic Books,

As both Plato and Heidegger were aware, Cupitt's observation has repercussions for the way we conceptualise "primary" and "secondary" conceptual orders. Preziosi writes:

> What artistry or artifice created for Plato, then, was not some 'second world' alongside the everyday world in which we live (the modern fantasy worlds of art history, museology, shopping, or digital media); he was quite clear that what art created *was* the world in which we really *do* live our daily lives. The problem he attempted to address was fundamental to philosophy, politics, and religion. If we believe that a particular made thing 'represents' some essence (either metaphorically 'contained' in some thing or absent and elsewhere – the 'soul' or 'spirit' of its time and place), then it is obvious that the essence purportedly 'represented' may be represented in other ways, problematizing the existence of that essence itself. Leading one to imagine that the essence supposedly represented is in fact created by its so-called 're-presentation.' Such an awareness obviously has the potential to undermine the claims of any political or religious power to security and truth. As Plato was perfectly aware in *The Republic* in his attempt to describe what would constitute an ideal state.
>
> It is this conundrum that is precisely the problem. The 'divine terror' (*theios phobos*) that art, according to Plato, induced in the soul was simply the terrifying awareness of exactly this: that works of art don't simply 'imitate' but rather *create and open up* a world, and keep it in existence [. . .]. This is an issue of the truth or falsity of imitation or representation: is a work of artistic creativity an imitation of some ideal essence, immutable truth, or transcendent reality, or is mutability itself what is 'transcendent'? Is the transcendent a supreme 'entity' above all things, or is it the taking-place, the very palpable re-

1997), 25.

> ality, of every thing? What, in other words, is the ontological status of an art object (physical or virtual)? Is it an effect of a pre-existing soul, spirit, or mentality, or is that spirit or mentality an after-effect of artifice itself? The materialism / immaterialism double-bind infects the heart of all of these problems.[27]

Theorists as different as Giorgio Agamben and Raymond Tallis seek a way forward that might bring together the opposed *philosophical* and *historical* perspectives with which we began this discussion.[28] Giorgio Agamben posits the concept of "bare life" as "the decisive event of modernity" and the focus of contemporary politics. Whereas Aristotle distinguished between contemplative life (*bios theóréikos*), the life of pleasure (*bios apolaustikos*) and the political life (*bios politicos*), Agamben draws our attention to a form of life overlooked by this trinity, namely *zoé*, or animal life. The individual, rather than being associated with the conduct appropriate to public life and the political sphere, is equated with "reproduction and subsistence", with *zoé* and *oikonomos.*[29] Yet 'bare life' is not simply 'animal' or 'biological' life, but is, as Hollis explains, "'biological life' that has entered the public realm and is now located at the centre of politics."[30] What is decisive in this politicization of bare life – and even, in Agamben's view, signals "a radical transformation of the political-philosophical categories of classical thought"[31] – is precisely that *zoé* is no longer bound by the confines of the *oikos*, or home, but has entered and occupied a cen-

27. Donald Preziosi, "Afterword: Artifice and Interactivity," *media-N: Art in the Age of Technological Seduction*, v.02, n.03 (2006), accessed May 10, 2011, www.newmediacaucus.org/media-n/2006/v02/n033.
28. See, for example, Rayond Tallis, *I Am: A Philosphical Inquiry into First-Person Being* (Edinburgh: Edinburgh University Press, 2004); Giorgio Agamben, *Homer Sacer: Sovereign Power and Bare Life*, trans. Daniel Heller-Roazen (Stanford, CA: Stanford University Press, 1998); Giorgio Agamben, *The Open: Man and Animal*, trans. Kevin Attell (Stanford, CA: Stanford University Press, 2004).
29. Agamben, *Homer Sacer,* 4. See also Catherine Hollis, "The politics of 'bare life': life and the body as the focus of power in the work of Benjamin and Agamben" (paper for the Political Studies Association 57th Annual Conference, University of Bath, April 11-13, 2007). Available online, accessed June 8, 2007, http://www.psa.ac.uk/2007/pps/Hollis.pdf.
30. Hollis, "The politics," 8.
31. Agamben, *Homer Sacer,* 4.

tral position in the polis with the result that "the division between *zoé* and *bios* is transcended and both become 'bare life' leading to the 'radical transformation' of the ontological categories on which politics and philosophy are based."[32] Sovereignty is thus no longer a historically specific form of political authority, but is rather the essence of the political; sovereign power, accordingly, is concerned with the production of bare life as the threshold of articulation between nature and culture, *zoé* and *bios*. It is concerned with separating real life from merely existent life, political and human life from that of the non-human. In Agamben's argument, therefore, the political is the production and definition of the inhuman, and its corresponding human analogue.

The irony here is that such attempts to theorise humanity's relationship to the animal and the inhuman is articulated in language – arguably, humanity's single most defining characteristic. Between the erasure of human selfhood and technoscience's seemingly unbounded pursuit of mastery over living matter, the humanities' preoccupations with origins persist, as is illustrated in Deleuze's account of the significance of Job:

> Job challenges the law in an ironic manner, refusing all second-hand explanations and dismissing the general in order to reach the most singular as principle or as universal.[33]

Instead of such self-defeating pursuit of origins, what would a future-oriented theory be, one setting out from such points of departure as unconscious, non-human, or even *unthinkable* perspectives? What potential metaphors and disciplinary hybrids arise from an imagination in which such dichotomies as human/post-human, artificial/authentic are supplanted by paradigms of theoretical hybridity, temporal multi-valence, 'post-disciplinary' creativity, risk-taking, innovative, action-oriented practice? Freeman Dyson, at the conclusion of his review of Richard Holmes' "The Age of Wonder", provides one vision of such a revitalised practice:

32. Hollis, "The politics," 7.
33. Gilles Deleuze, *Difference and Repetition,* trans. Paul Patton (London: Continuum, 2004), 8.

> If the dominant science in the new Age of Wonder is biology, then the dominant art form should be the design of genomes to create new varieties of animals and plants. This art form, using the new biotechnology creatively to enhance the ancient skills of plant and animal breeders, is still struggling to be born. It must struggle against cultural barriers as well as technical difficulties, against the myth of Frankenstein as well as the reality of genetic defects and deformities.
>
> If this dream comes true, and the new art form emerges triumphant, then a new generation of artists, writing genomes as fluently as Blake and Byron wrote verses, might create an abundance of new flowers and fruit and trees and birds to enrich the ecology of our planet. Most of these artists would be amateurs, but they would be in close touch with science, like the poets of the earlier Age of Wonder.[34]

And so the pursuit (or critique) of origins – the source and evolution of an idea, method, or praxis – while no doubt interesting in itself, becomes somewhat less compelling, particularly at a time of increasing resource scarcity, than the creation of new knowledge through transdisciplinary co-operation and action, supported by transverse epistemologies[35]. We may pursue such forms of knowledge by cross-pollinating visions, methods, speculative prototypes – and then applying our insights to create *the difference that makes the difference*.

34. Freeman Dyson "When Science and Poetry were Friends," *The New York Review of Books* (August 2009). Review of *The Age of Wonder: How the Romantic Generation Discovered the Beauty and Terror of Science* by Richard Holmes, accessed February 8, 2011, http://www.nybooks.com/articles/archives/2009/aug/13/when-science-poetry-were-friends/?page=1.
35. See Rolf Hughes, "The Art of Displacement: Designing experiential systems and transverse epistemologies as conceptual criticism," *Footprint*, Issue # 4, (Spring 2009): 49-63. See also Rolf Hughes and Ronald Jones: "Modern 2.0: Post-criticality and transdisciplinarity" in *Transdisciplinary Knowledge Production in Architecture and Urbanism: Towards Hybrid Modes of Inquiry,* ed. Nel Janssens and Isabelle Doucet (Dordrecht: Springer, 2011).

Cryptobiologies

Eugene Thacker

There is a great deal of code-making and code-breaking in biotechnology. We "crack" the genetic code, "decode" the genomes of various organisms, "encode" those codes into actual computer databases, all to help us decipher the information of disease-causing agents, which themselves are able to evade medicines by their rapid rate of genetic mutation. Yet, in the midst of all this talk of codes, we often forget that many of the applications of industrial biotechnology result not in codes but the flesh of life: mice, sheep, pigs, goats, and so on. Their use in livestock breeding, transgenics, and medical research suggest to us that we have not only decrypted the code of life, but we have advanced to a level where we can encrypt life in the form of these unique animals.

However our relationship to animals is at best a complicated one. The history of Western thought on the topic can be viewed as a continued effort to separate the human from the animal (Aristotle's description of man as a "political animal," Descartes' formulation of the *bête machine*, the debates surrounding *The Descent of Man*). The search for the set of characteristics that would definitively separate human from animal often presumes a clear division between the natural and the artificial, or what today we would refer to as biology and technology. Yet, even a cursory look at biotechnology today suggests that something is afoot. What happens when we produce animals that are not "natural"? What do we make of these biologies that are also technologies? Are they of nature, of technology, or of something else entirely? How do we relate to these non-natural, even super-natural animals?

What I would like to do here is to briefly present three cultural relationships between human and animal, relationships that not only challenge us to rethink the animal, but also the human. In an everyday sense, we coexist with animals of all types, from our domesticated dogs and cats to the animals displayed in the meat, poultry, and seafood sections of the grocery. We call to animals, and we also eat animals. We develop, with our pets, unique modes of communication, and, with our food, we also develop unique modes of consumption. In this everydayness of the animal, in this quotidian relation we have with animals, we as human beings practice this dual form of orality – communicating and consuming, speaking and eating, word and flesh.

But what of animals that are not everyday? What of human-animal relationships that are far from ordinary, but are rather extraordinary? Of course, exotic animals can also be pets, in which case the exotic becomes everyday. So perhaps a better question to ask is, are there instances in which the human-animal relationship occupies a grey zone in between the everyday and the exceptional, the ordinary and the extraordinary?

Biotech Animality

Genetic engineering as applied to animals occupies a curious position in Western, technologically-advanced cultures. It is at once the most hi-tech and esoteric method of working with nature, and yet its applications are the most quotidian (food, pets). Certainly breeding techniques have been known for many, many years, and their applications in domestication and farming have been documented by archaeologists, anthropologists, and historians. However the introduction of genetic engineering techniques into the biotech industry in the 1970s has had a profound impact on the way we view the human-animal relationship – an impact we are undoubtedly still witnessing. To take a few, well-known examples: genetically-modified organisms (GMOs), which, in the broadest sense, may be taken to include microbes (e.g. bacteria that digest oil spills), the whole range of cloned mammals in science research (cloned sheep, mice, cows, pigs, monkeys), the field of transgenics (e.g. goats genetically engineered to produce human insulin in their milk), biotech livestock (meatier chickens, fatter pigs, etc.), and of course genetic engineering applied to domestic pets (e.g. allergen-free cats).

These and other examples constitute our contemporary biotech bestiary, a whole new natural history of the biotech zoo, a whole new classification system of previously impossible creatures, hybrids and teratologies that would seem to be the domain of fantasy more than fact. Certainly science fiction itself often speculates on the possibilities of such impossible beings, but what is equally fascinating is the moment in which such seemingly impossible biologies cross a certain threshold and become everyday technologies. Our perplexity in attempting to comprehend the very existence of GMOs, transgenic animals, cloned mammals, and genetically engineered pets is an indicator of the grey zone occupied by this biotech bestiary.

Like the mediaeval bestiary, our contemporary biotech bestiary is filled with beings that exist in a direct relation to an epistemological structure, a sort of hermeneutical *topos* of physiology, natural behavior, and cultural meaning. But this was not an exclusive concern of the Middle Ages; the bestiaries themselves drew upon earlier examples, including the fourth-century *Physiologus*, as well as the natural histories of Pliny and works of Aristotle such as the *Historia Animalium*. Here a concern with knowledge-production about the natural world was combined with a fascination for ordering and classifying living beings. However, it was with the Medieval bestiary that these ordered and named beings also took on explicit moral and theological meanings, though the range of classical and mythical symbology never completely disappeared from such texts. But its primary purpose was essentially to theologize through the vernacular provided by the natural world – to, in effect, speak the divine *logos* through the beast. Often this was done by first offering a description of a beast, and then going to elaborate its deeper moral or theological meanings:

> The name "beast" applies, strictly speaking, to lions, panthers, and tigers, wolves and foxes, dogs and apes, and to all other animals which vent their rage with tooth or claw – except snakes. They are called "beasts" from the force with which they rage. They are called "wild" because they enjoy their natural liberty and are borne along by their desires.[1]

As one of the authors of the Aberdeen Bestiary states, the purpose of such books was "to improve the minds of ordinary people, in such

1. *The Book of Beasts: Being a Translation from a Latin Bestiary of the Twelfth Century*, ed. and trans. T.H. White (Mineola: Dover, 1984), MS 24, f7r.

a way that the soul will at least perceive physically things which it has difficulty grasping mentally."[2] Isidore of Seville's *Etymologiae* includes a book devoted to creaturely life, and in so doing connects life and word, beast and spirit, under the creation of God. This impetus was further formalized in the most popular existing bestiaries, such as the one at the University of Aberdeen, which is now online. The bestiaries were collaboratively authored and, in many cases, simply lifted or borrowed from earlier versions (of which the *Physiologus* seems to have been the most popular source).

The bestiary is also related to another genre of animal books, which is the teratology – the set of all animals that have no set. By definition, the impossible animal, the fantastic being, the monster, are all forms of unnatural life, or even life that cannot – or should not – exist. But more than this, the monster also throws up a challenge to the very concept of nature and of our relation to and distance from that which we call natural. From the early modern era to the 19th century, the study of monsters (derived from the Latin *monstrum* – "to warn") is this attempt to comprehend the animal being that does not fit, the animal life that has no home, no proper place. Teratology – the study of monsters – is a documentation of this animal displacement. From Ambrose Paré's *Des Monstres et Prodiges* (1573) to Geoffroy Saint-Hilaire's *Histoire Générale et Particulière des Anomalies, ou, Traité de Tératologie* (1832), the treatise on monsters is, in a sense, a classification of unnatural life or life that should not exist. Such studies are positioned between naturalistic explanations of anomalies and a range of supernatural interpretations. Monsters oscillate between being divine prophecies, a display of the wonders of nature, and medical-scientific errors deviating from a norm.

In his delightful *Book of Imaginary Beings*, the Argentinean author Jorge Luis Borges discusses our dual fascination with the real animal kingdom and with the impossible animals that inhabit myth and folklore: "Let us now pass from the zoo of reality to the zoo of mythologies, to the zoo whose denizens are not lions but sphinxes and griffons and centaurs. The population of this second zoo should exceed by far the population of the first, since a monster

2. *The Book of Beasts*, f25v.

is no more than a combination of parts of real beings, and the possibilities of permutation border on the infinite."[3] Borges compiled his book prior to the era of genetic engineering, but it is tempting to read his comments on hybrids and recombination in relation to our current biotech bestiary. We might even wonder if there exists a whole "micro-monstrosity" of viruses, bacteria, fungi. This is the term used by philosopher of science Georges Canguilhem, who, a few years after the Watson-Crick publications, wondered if the historical interest in teratologies has been transformed into a current concern with information, noise, and error.[4]

Becoming-Microbial

Surely we as human beings are more than the microbes that inhabit our bodies and that sustain many of our biological processes. Microbes, strictly speaking, are not animals – they're microbes. We are animals...we think – except that our thinking about our animality makes us more than animals. Yes (we say to ourselves), we are more than our microbes. Except, of course, when our microbes are not ours (infection), or when our microbes are always coming-and-going (contagion). The biological processes of contagion and infection always elicit a certain anxiety and fear for us, and for good reason. Contagion and infection are more than mechanisms of antigen recognition and antibody response; they are, as our textbooks tell us, entire "wars" and "invasions" continuously fought on the battle lines of the human body (to which autoimmune disorders add degrees of metaphorical complexity).

Contagion and infection are paradoxical processes. They elicit a rigorous "defense" of the body's boundaries, and yet we as living beings are defined by our continuous exchange of matter and energy with our surroundings. Only certain things are allowed to pass, only certain things are exchanged. All of this denotes a systems-wide, *network* perspective. It is no accident that computer networks, economic exchanges, and cultural ideas have been described in terms of viruses (computer viruses, viral marketing, me-

3. Jorge Luis Borges, *The Book of Imaginary Beings*, trans. Norman Thomas di Giovanni (New York: Penguin, 1974), 14.
4. Georges Canguilhem, *The Normal and the Pathological*, trans. Carolyn Fawcett (New York: Zone, 1991), 275ff.

mes). There is an abstract topology, a network form, that pervades each of these systems. They are constituted by nodes and edges (dots and lines) that have variable rates of exchange and connectivity. Such networks have several forms, or topologies, each with an analogous control structure: centralized, decentralized, and distributed. It is for this reason that many network science perspectives have studied biological and computer viruses interchangeably: the microbe is the "message" that is passed along channels of contagion (the edges) between each person (the nodes).

Thus, the "war" that takes place in contagion and infection is not simply limited to the body's interior; it is also a conflict that is scaled up, as it were, to the level of the population, and indeed, the nation. This is the point where virology and immunology fold onto epidemiology and public health. The task of public health agencies is thus to distinguish "good" circulations (travel, trade) from "bad" circulations (virulent microbes). What public health organizations such as the WHO and the CDC call emerging infectious diseases are networks in this way. Microbes establish networks of infection within a body, and networks of contagion between bodies, and our modern transportation systems extend that connectivity across geopolitical borders ("global health").

However, it is misleading to say that microbes "do" this or that they "do" that, as if they were little homunculi with malintent. But it is equally misleading to simply say that we humans do this or do that, especially as most epidemics involve many factors that include microbial evolution, drug-resistance, and environmental factors, in addition to the more human concerns of education, preventive practices, and prescription drugs. Indeed, if microbes are in some way synonymous with networks, then the whole question of agency is rendered problematic. It is this that incites the greatest discomfort. How is it started? How can it be stopped? How can it be prevented? Not only do the networks of contagion and infection render human agency and control problematic, but, when we take into account all the factors that go into an epidemic, we see as many nonhuman agencies as human ones (e.g. viral mutation, bacterial resistance). Representations of epidemics in popular culture – from Daniel Defoe's *A Journal of the Plague Year* to contemporary zombie films such as George Romero's *Land of the Dead* – can be understood as cultural reactions to this strange, fearful, "nonhuman life" of microbial networks.

In fact, we are still unsure as to whether viruses are living or non-living – they seem to be simple assemblages of matter without the ability to independently reproduce, and yet recent research has revealed their troubling ability to genetically mutate and exchange genetic material with a host organism. Virologists such as Luis Villareal (echoing the work of Lynn Margulis) have suggested that the old question of the living/non-living status of viruses is superceded by another question: the role that viruses have played in evolutionary processes, whether or not they are "alive." It seems that microbes are not only very, very old, but that they have developed innovative ways of living with (and inside) we human beings. Should we say the reverse as well, that human beings have developed innovative ways of living with microbes?

Whatever Life

One of the hallmarks of 21st century U.S. biodefense policies has been the implosion between emerging infectious disease and bioterrorism, a collapse of a distinction in cause in favor of a unity in effect. Nowhere is this more evident than in the conceptual – even ontological – articulations performed in the language of biodefense. For instance, the U.S. 2002 Bioterrorism Act contains at numerous points a refrain, one that can also be heard in other national and homeland security documents: "bioterrorism *and* emerging infectious disease." The opening sections of the Bioterrorism Act give public health administrators the ability to develop strategies "for carrying out health-related activities to prepare for and respond effectively to bioterrorism and other public health emergencies, including the preparation of a plan under this section."[5] Here, the word "and" plays a central role in the document as a whole, implying a certain quality of *whatever*: the notion that "bioterrorism and emerging infectious disease, it makes no difference which," that is also a notion of "whichever it is, it matters a great deal."[6]

5. Title XVIII, Subtitle A, Section 2801. The full title is *Public Health Security & Bioterrorism Preparedness & Response Act of* 2002.
6. Giorgio Agamben, *The Coming Community*, trans. Michael Hardt (Minneapolis: University of Minnesota Press, 2003), 1.

However, the most remarkable consequence of this implosion is in what the "and" enables in the way of public health practices. As part of a broad endeavor to facilitate biodefense research, the U.S. Project BioShield has, since its announcement in 2002, allocated funding for the development of "next-generation medical countermeasures" such as drugs, vaccines, and diagnostics. In 2003 the U.S. National Institute of Allergy and Infectious Disease (NIAID), a department within the National Institute of Health (NIH), received a multi-million dollar award for research into "human immunity and biodefense." Later that same year, NIAID officials released a progress report outlining their research goals. The report states that the "increased breadth and depth of biodefense research not only is helping us become better prepared to protect citizens against a deliberately introduced pathogen, it also is helping us tackle the continuous tide of naturally occurring emerging infections."[7] Distinctions in cause are effaced by the biological latency of the disease-causing agent, a latency that is also social, political, and economic – precisely because it is biological. Indeed, it is this notion – that biology is more-than-biological because it is biological – that can be said to be the conceptual foundation for the flurry of 21st century U.S. biodefense legislation: The Bioterrorism Act, Project BioShield, the Biosurveillance Project, the National Electronic Disease Surveillance System (NEDSS), the National Pharmaceutical Stockpile, as well as a host of classified projects.

However, we can note a more fundamental issue at stake in these developments, and this surrounds the problematic of biological "life itself." By this phrase I mean the ways in which the domain of the biological – a shifting and discontinuous domain, to be sure – is articulated as a problem of control, regulation, and modulation, a condition that Michel Foucault has described as biopolitical.[8] The problematic of biological life itself also denotes the ways in which the domain of the biological is rendered as tech-

7. U.S. National Institute of Allergies and Infectious Disease (NIAID), "NIAID Biodefense Research Agenda for CDC Category A Agents – Progress Report" (29 September 2003).
8. The phrase 'life itself' refers to a concept employed by molecular biology researchers in the 1950s and 1960s (foremost among them François Jacob), as well as its more critical use in science studies by Melinda Cooper, Robert Mitchell, and Nikolas Rose.

nically specific (in viruses, bacteria, genomes, vaccines) as well as a pervasive, general, even existential condition (the presumed facticity or givenness of life itself). For Heidegger, one of the ways in which *Dasein* or Being reveals itself is in the *Angst* associated with the very fact of existence. This *Angst* is to be differentiated from the fear of particular things and the particular threat they represent; thus *Angst* is not fear. "That about which one has *Angst* is being-in-the-world as such...What *Angst* is about is not an innerwordly being...The threat does not have the character of a definite detrimentality which concerns what is threatened with a definite regard to a particular factical potentiality for being. What *Angst* is about is completely indefinite."[9]

Except – and this is the crucial difference – Heidegger's distinction revolved around the question of *Dasein*, and not the question of biological life itself. In fact, for Heidegger, the question of life was not a question at all, for the sciences of biology and psychology, in their asking of the question "what is life?" mistakenly presume to have already answered the more fundamental question "what is Being?"[10] However, while Heidegger dismisses the question of biological life itself, what we are witnessing in the ontology of biodefense is a certain conceptual displacement. Whereas Heidegger contrasted the question of Being (in terms of *Angst*) with the question of life (as fear), today we have a reformulation of the latter in terms of the former – an *Angst* that is about biological life itself. In biodefense, *Angst* is correlated to biological life itself. That about which one has Angst is the pervasiveness of the biological as threat, as what is threatened, and as response. "The fact that what is threatening is *nowhere* characterizes what *Angst* is about."[11] The logic of biodefense – that life itself is an indefinite and indeterminate threat – culminates in a social, cultural, and political *Angst*, a biological *Angst*, an *Angst* of life itself. Here the problematic of life itself is how to articulate, within the domain of the living, that which is threatening versus that which is threatened, resulting in as peculiar type of existential biology.

9. Martin Heidegger, *Being and Time*, trans. Joan Stambaugh (Albany: State University of New York Press, 1996), §40, 174.
10. Heidegger, *Being and Time*, §10, 42ff.
11. Ibid., 174.

Occult Biologies

If contagion and infection can be seen as networks, and if such networks incite fear in us, in part due to their nonhuman character, how do we comprehend this ambivalent, affective dimension to biological life? Writing about the politics of public health responses to disease, Michel Foucault notes that plagues have historically elicited two responses: a "poetic fantasy of lawlessness" (social anarchy, the "dance of death") and a "political fantasy of total control" (quarantines, pesthouses, death tables).[12] Foucault's comments ask us to view contagion and infection as being more-than-biological – as social, cultural, and political as well.

A historical look at epidemics reveals this aspect of the more-than-biological. For instance, epidemics are often found where there is war or military conflict. Thucydides remarks that, during the Peloponnesian War, there were rumors of the wells being intentionally poisoned – a possible early example of biological warfare. The medieval practice of catapulting diseased and/or decaying cadavers of soldiers and animals would carry this further. The Great Plague of London in 1665 took place in the midst of civil war, and it was no accident that Thomas Hobbes would compare civil dissent with a "diseased" body politic in his *Leviathan*. Epidemics are not only found in the midst of war, but, they are often interpreted in ways that are more than medical or natural. During the Black Death, which ravaged most of Europe in the mid-14th century, the predominant explanations were, unsurprisingly, religious. Italian and German chroniclers of the period note the predominance of religious processionals, flagellant groups, and the apocalyptic exhortations of popular soothsayers. In the era of European expansionism, disease – which often accompanied imperial and colonial enterprises – was often interpreted by both colonizer and colonized as a sign of divine retribution or providence, depending on the point of view.

It is with scientific hindsight that we have since demystified such supernatural interpretations of epidemics: the plague bacillus, we are told, was carried in fleas, living on rats, themselves pop-

12. Michel Foucault, *Discipline and Punish*, trans. Alan Sheridan (New York: Vintage, 1995), 195ff.

ulous aboard merchant ships traveling between southern Europe and the Mongol region. But an exclusive reliance on scientific facts – however useful – obscures the ambivalent, affective cultural dimensions of epidemics. The bacillus-flea-rat connection is perhaps culturally reflected in religion, myth, folklore – from the Brothers Grimm modernization of the Pied Piper of Hamelin to Werner Herzog's expressionist tribute *Nosferatu* – there is an entire cultural history of plague that has yet to be written. Such a history would have to feature animals, not just as carriers of disease, but as carriers of disorder, filth, impurity – even as carriers of divine retribution. Rats, bats, and packs. There are always many of them; it is rarely a single rat, a single flea, a single bacillus that is the harbinger of disease. Gilles Deleuze and Félix Guattari note that there are three types of animals: anthropomorphic, domesticated pets (the mirror of the human), our scientific species (official, institutional, "state animals"), and finally there is a third type of animal, the pack or the swarm animals, the animals that do not exist except as many – animal multiplicities.[13] They are not "a" bee, but the swarm, not "a" bird but the flock, not "a" bacterium but the pandemic. This third animal is traditionally interpreted as an underworld animal, an animal without head or face, a demonic animal – "I am legion."[14]

Weird Biology

We return again to the question of the animal – or rather, of animality. In the case of emerging infectious disease, animals as groups often become the links between human and human (mad cow, monkeypox, bird flu, etc.). But beneath this is another level of animality, that of microbes passing between organisms, microbes exchanging genetic material in networks of contagion and infection. Is this too an instance of animality? In modern fiction, the under-appreciated genre of supernatural horror is replete with examples of a contagious, swarming life that is also radically

13. Gilles Deleuze and Félix Guattari, *A Thousand Plateaus*, trans. Brian Massumi (Minneapolis: University of Minnesota Press, 1987), 26ff.
14. This idea is further explored in my essay "After Life: Swarms, Demons, and the Antinomies of Immanence," in *Theory After Theory*, edited by Jane Elliott and Derek Attridge (New York: Routledge, 2011), 181-93.

non-human and unnatural – H.P. Lovecraft's ancient, formless Shoggoths, Clark Ashton Smith's primordial, amorphous Ubbo-Sathla, Frank Belknap Long's surrealistic Hounds of Tindalos and the entire dark matter bestiary of William Hope Hodgson's *The Night Land*. For this reason, formless, pack, or swarm animals – even when presented as epidemics – show us an animality that we *apprehend* but do not *comprehend*. Philosopher of the formless Georges Bataille reiterates this: "The animal opens before me a depth that attracts me and is familiar to me. In a sense, I know this depth: it is my own. It is also that which is farthest removed from me, that which deserves the name depth, which means precisely *that which is unfathomable to me*."[15] And our apprehension of such animals is ambivalent, precisely because they symbolize radical, non-human transformations. This is why supernatural explanations predominate in historical instances of plague, and this is also why the genre of supernatural horror is the domain in which we find nameless offspring and logical monsters.

To say that we as human beings cannot really know what it is like to be an animal would be commonplace. But to ask what it would be like to be a pack, a swarm, a flock – this is the question of animality. It is a more abstract question, a question not of species, genus, and organism, but of topologies or patterns that effortlessly cut across species. The threshold of our understanding is not between human and animal, but rather between humanity and animality. As Jorge Luis Borges notes, "we are ignorant of the meaning of the dragon in the same way that we are ignorant of the meaning of the universe, but there is something in the dragon's image that fits man's imagination, and this accounts for the dragon's appearance in different places and periods."[16]

15. Georges Bataille, *Theory of Religion*, trans. Robert Hurley (New York: Zone, 1992), 22.
16. Borges, *The Book of Imaginary Beings*, 12.

Contributors

JAY DAVID BOLTER holds the Wesley Chair in New Media and is the Co-Director of the Wesley Center for New Media Research and Education. His publications include *Writing Space: Computers, Hypertext, and the Remediation of Print* (Lawrence Erlbaum and Associates, 1991/2001), *Remediation: Understanding New Media* (with Richard Grusin, MIT Press, 1999) and Windows and Mirrors: Interaction Design, Digital Art and the Myth of Transparency (with Diane Gromala, MIT Press, 2003).

RON BROGLIO is an Assistant Professor of English at Arizona State University. He is author of *Technologies of the Picturesque* (Bucknell UP, 2008) and *Surface Encounters: thinking with animals and art* (Minnesota University Press, 2011)

BOO CHAPPLE is an artist and researcher who writes and exhibits work in diverse media. Her work has been exhibited at Ars Electronica, the Beijing Biennale of Architecture, and SF MoMA. Her writing has been published in Art of the Biotech Era (Milentie Pandilovski, ed.) and Plastic Green (Pia Ednie-Brown, ed.). Chapple holds a Masters of Design by research from RMIT University, Melbourne and has taught at universities in Australia and The Netherlands.

MARIA CHATZICHRISTODOULOU [a.k.a. MARIA X] is a cultural practitioner (curator, performer, producer, writer), Director of Postgraduate Studies and Lecturer in Theatre and Performance at the School of Arts and New Media, University of Hull. She holds a PhD in Art and Computational Technologies from Goldsmiths Digital Studios, University of London. She was co-founder and co-curator of the international festival Medi@terra (Athens, Greece, 1997-2002), co-diector of the festival and symposium Intimacy: Across Visceral and Digital Performance (London, 2007) and curator of the Expanded Play events (Zaragoza, 2010 and ongoing).

She is a co-editor of the volume *Interfaces of Performance* (Ashgate, 2009) and the forthcoming volume *Intimacy Across Visceral and Digital Performance* (Palgrave Macmillan).

MARTINE ÉPOQUE has been professor in the Département de danse from 1980 to 2008, and is now professor emeritus at Université du Québec à Montréal. A prolific choreographer whose stage and screen works have been met with international acclaim, she has received the Clifford E. Lee choreographic award in 1983 and the Prix du Québec Denise-Pelletier (interpretative art) in 1994. She founded the Laboratoire d'application et de recherches en techno-chorégraphie (LARTech) in 1999.

ROLF HUGHES is Professor in Design Theory and Practice-Based Research at Konstfack University College of Arts, Craft and Design, where he teaches on the MFA courses in Experience Design, Ceramics and Glass, and Storytelling. He is also Senior Professor at the Sint-Lucas School of Architecture (Brussels & Ghent, Belgium), where he has been developing an innovative practice-led PhD programme for the past five years, and a member of faculty for the Stockholm School of Entrepreneurship (www.sses.se). In 2010, Hughes was elected onto the expert committee for artistic research at Vetenskapsrådet, the National Research Council of Sweden, and in 2011 he was elected Vice President for the international Society for Artistic Research. A published prose poet, he holds a PhD in Creative and Critical Writing from the University of East Anglia (1994).

FRANÇOIS-JOSEPH LAPOINTE is full professor in the Département de sciences biologiques at Université de Montréal, as well as PhD candidate in arts at Université du Québec à Montréal. As an artist and researcher in bioart, he is interested in the application of biological concepts and genetic algorithms to dance composition. He is the current director of the Laboratoire d'écologie moléculaire et évolution (LEMEE).

JOHN MONK is Emeritus Professor at the Open University where he has over 35 years experience of teaching at a distance and researching in electronics and digital systems. His research activities have primarily been industry-based. He is heavily involved with professional engineering bodies in the registration of professional engineers and the establishment of professional standards and has a strong interest in engineering ethics and the relationship between philosophy and technology.

JENNY SUNDÉN is Associate Professor at the School of Gender, Culture and History, Södertörn University, Sweden. Her work is situated in the areas of new media studies, cultural studies, science and technology studies, queer/feminist theory, and games. She is the author of *Material Virtualities: Approaching Online Textual Embodiment* (Peter Lang 2003), and a co-author of *Gender and Sexuality in Online Game Cultures: Passionate Play* (Routledge forthcoming, with Malin Sveningsson).

EUGENE THACKER is Associate Professor in the Media Studies Program at the New School, New York. He is the author of *After Life*, published by the University of Chicago Press.

KARIN WAGNER has a PhD in art history and visual studies from Gothenburg University and works as a senior lecturer at Chalmers University of Technology, Gothenburg. Her research interests include photography, new media and visual communication. She is currently working on the project "The (Un)sustainable Package" funded by the Swedish Research Council.

TIMOTHY WEAVER is Associate Professor of eMAD and Digital Media Studies at the University of Denver where his research areas include new media-based biological narrativity, biomimetics, ecosemiotics, and sustainable design. As a new media artist and former life scientist his concerted objective has been to contribute to the restoration of ecological memory through speculative inquiry across the art | science interface. His recent video, sound and installation projects have been featured at over 80 venues in North and South America, Europe and Asia.

www.ingramcontent.com/pod-product-compliance
Ingram Content Group UK Ltd.
Pitfield, Milton Keynes, MK11 3LW, UK
UKHW021705190726
13853UKWH00001B/430